T0245262

CAMBRIDGE LIBRARY COLLECTION

Books of enduring scholarly value

Life Sciences

Until the nineteenth century, the various subjects now known as the life sciences were regarded either as arcane studies which had little impact on ordinary daily life, or as a genteel hobby for the leisured classes. The increasing academic rigour and systematisation brought to the study of botany, zoology and other disciplines, and their adoption in university curricula, are reflected in the books reissued in this series.

Memoirs of the Botanic Garden at Chelsea

Henry Field (1755–1837) was a British apothecary and member of the Society of Apothecaries of London. Besides serving in various administrative capacities for the Society, as well as for the London Annuity Society (founded by his father), he was nominated in 1831 as one of the medical officers for the City of London board of health, charged with taking precautions against an outbreak of cholera in the city. A lecturer and regular contributor to medical journals, Field is also the author of this history of the Chelsea Physic Garden, first published in 1820. The present reissue, published in 1878, was revised and extended by Robert Hunter Semple (1815–91). The garden was originally created by the Society as a professional resource in 1673 and the book covers its development up to 1878, and also includes a ground plan of the garden in that year.

Cambridge University Press has long been a pioneer in the reissuing of out-of-print titles from its own backlist, producing digital reprints of books that are still sought after by scholars and students but could not be reprinted economically using traditional technology. The Cambridge Library Collection extends this activity to a wider range of books which are still of importance to researchers and professionals, either for the source material they contain, or as landmarks in the history of their academic discipline.

Drawing from the world-renowned collections in the Cambridge University Library, and guided by the advice of experts in each subject area, Cambridge University Press is using state-of-the-art scanning machines in its own Printing House to capture the content of each book selected for inclusion. The files are processed to give a consistently clear, crisp image, and the books finished to the high quality standard for which the Press is recognised around the world. The latest print-on-demand technology ensures that the books will remain available indefinitely, and that orders for single or multiple copies can quickly be supplied.

The Cambridge Library Collection will bring back to life books of enduring scholarly value (including out-of-copyright works originally issued by other publishers) across a wide range of disciplines in the humanities and social sciences and in science and technology.

Memoirs of the Botanic Garden at Chelsea

Henry Field
Edited by R.H. Semple

CAMBRIDGE UNIVERSITY PRESS

Cambridge, New York, Melbourne, Madrid, Cape Town,
Singapore, São Paolo, Delhi, Tokyo, Mexico City

Published in the United States of America by Cambridge University Press, New York

www.cambridge.org
Information on this title: www.cambridge.org/9781108037488

This edition first published 1878
This digitally printed version 2011

ISBN 978-1-108-03748-8 Paperback

SIR HANS SLOANE, BART., M.D.

(From an Engraving in the British Museum.)

MEMOIRS

OF THE

BOTANIC GARDEN AT CHELSEA

BELONGING TO THE

Society of Apothecaries of London.

BY THE LATE HENRY FIELD, ESQ.

REVISED, CORRECTED, AND CONTINUED TO THE PRESENT TIME,

BY R. H. SEMPLE, M.D.

WITH ILLUSTRATIONS, PLANS OF THE GARDEN, AND A CATALOGUE OF THE PLANTS, ARRANGED ACCORDING TO THE NATURAL SYSTEM.

LONDON:

PRINTED BY GILBERT AND RIVINGTON,

ST. JOHN'S SQUARE, CLERKENWELL.

1878.

TO THE

MASTER, WARDENS, COURT OF ASSISTANTS,
AND MEMBERS

OF THE

Society of Apothecaries of London,

THESE

MEMOIRS OF THE CHELSEA GARDEN

ARE MOST RESPECTFULLY INSCRIBED

BY

THE EDITOR.

VIEW OF CHELSEA GARDEN FROM THE RIVER, BEFORE THE CONSTRUCTION OF THE CHELSEA EMBANKMENT.

MEMOIRS

OF THE

BOTANIC GARDEN AT CHELSEA.

Of the various and important events recorded to have taken place, during the two centuries and a half which have elapsed since the Society of Apothecaries became a separate and independent Corporation, there is none which reflects greater honour upon it as a Society, and more forcibly evinces the zeal, energy, and disinterestedness of its early members in the promotion of science, than the establishment of a Botanic Garden at Chelsea.

Were not the fact indisputable, it would scarcely be credited by posterity that this expensive design was commenced at a time when the Society could not with propriety be said to have emerged from its state of infancy; when it was totally destitute of any disposable funds which could be employed for that purpose; and at a period when their Hall had been recently destroyed by the memorable fire of London in 1666. The early re-edification of their Hall, so necessary for the conduct of their general affairs, was indispensable, and must have been a primary object with the members. Their private pecuniary resources would have been, according to every rational conjecture, too much exhausted by this necessary work to enable them to enter upon a new undertaking, the only design of which was honourable reputation without any prospect of worldly advantage.

The earliest record we possess of a Botanic Garden in England was that of the celebrated John Gerarde, the father of English herbalism. He flourished in the latter part of the 16th, and the beginning of the 17th Century. His garden was attached to his house in Holborn, and he published, in the years 1596 and 1599, a catalogue of the plants it contained.

The next Botanic Garden in order of time was that of a man equally celebrated, though in a somewhat different way, namely, John Tradescant, gardener to King Charles I., and who, about the year 1630, established at South Lambeth a garden for the cultivation of exotic plants, a project which was carried out with great success for many years by himself and his son. The name of Tradescant survives in the world of science in the designation given to the pretty plants called *Tradescantia*, or Spiderworts. Tradescant, besides being a good botanist, was a great collector of objects in other branches of Natural History; and his collection, which is a very valuable one, was left to Mr. Elias Ashmole, by whom it was, in 1677, presented to the University of Oxford, where it still exists in the well-known Ashmolean Museum of that ancient seat of learning.

Mr. William Watson, a member of the Apothecaries' Society, afterwards Sir William Watson, M.D., visited this garden at South Lambeth in 1749. He informs us that Mr. Tradescant's garden had been for many years totally neglected, and that the house belonging to it was now empty and ruinous, but that, although the garden was quite covered with weeds, there were still to be seen manifest traces of its founder. He enumerates the names of a few plants, and he speaks of two *Arbutus* trees of a very large size, and of a *Rhamnus catharticus*, about twenty feet in height and nearly a foot in diameter, all of which still remained in the garden.

The next Botanic Garden in succession was most probably that of the Society of Apothecaries, which in a few years arrived at a very honourable degree of eminence, and continued to flourish with increasing energy and success during a long period, while its predecessors, above mentioned, had fallen into decay and had left scarcely a trace of their former existence.

Collections of plants, both native and exotic, in the gardens of private gentlemen and of scientific nurserymen, were no doubt at all times numerous and extensive, but the Society of Apothecaries for a lengthened period might justly claim the honour of possessing the only large garden devoted to Medical and general Scientific Botany in the vicinity of the metropolis, and supported by a public body.

The advancement of Botany in its relation to Medicine, although the principal, was by no means the only object of the founders of the Society's Garden at Chelsea, who also aimed at introducing the vegetable productions of foreign climates into this country, and of placing them in such conditions, either natural or artificial, as might be necessary to acclimatise them and to present them, in the most favourable aspects and in the most enduring vigour, to the notice of botanical students. A further and no less useful object has always been kept in view by those who have maintained the Garden and have superintended its administration, namely, to classify as far as possible and in the most convenient manner, the multitudinous individual forms which constitute the Vegetable Kingdom, and so to render the collection at Chelsea a well-arranged and systematic epitome of most of the groups of plants which are spread over the surface of the globe.

The last-named object aimed at in the management of the Chelsea Garden is of course difficult of attainment, and the modes in which it has been carried out have necessarily varied at different epochs. To discover a method of grouping, in distinct and definite subdivisions, the great host of vegetable productions of the earth and the sea is almost to penetrate into the mysterious secret of the plan of creation; and it cannot be a matter of wonder that such a task, although it has occupied the minds of the greatest philosophers from the dawn of civilisation down to the present day, has never been attended with complete success. Nevertheless, sufficient materials have been collected, and an adequate amount of conclusions have been already drawn, to demonstrate the beautiful simplicity, amidst apparently infinite diversity, which prevails in the vegetable world as in all other departments of Nature, and the successive labours of Ray, Tournefort, Linnæus, Jussieu, and others have not been in vain in unfolding to the admiration of mankind the wonderful uniformity of the elementary structures of plants and the ineffable skill with which these structures have been wrought into myriads of shapes. All these shapes are linked together by alliances and affinities which it is the province of the philosophic botanist to discover and to describe, and the SYSTEMS, as they are called, of Botanical Classification are the feeble expressions of the human intellect to unravel the laws which govern the distribution of vegetable forms. As it is the duty of the scientific musician to resolve a harmony into its component parts or notes, so it is the aim of the botanist to resolve the harmonies of the vegetable world into those primitive types or forms, the boundless varieties and modifications of which constitute the verdant carpet of the earth.

The classification of plants, then, has been one of the main objects held in view by the founders and managers of the Chelsea Garden, and a large portion of the ground has long been, and still is devoted to the purpose of classical and scientific arrangement, apart from and independent of the general distribution of the plants for ornamental or useful purposes.

From the remarks above made it will be readily understood why the principles of classification have varied in the Garden in different periods, for although the laws of Nature always remain the same, human systems are continually changing. Thus in the seventeenth century the crude but ingenious systems of Ray and Tournefort prevailed at Chelsea, to be succeeded towards the end of the eighteenth by the more precise but artificial classification of Linnæus, which in its turn was to give way, towards the middle of the nineteenth century, to the NATURAL SYSTEM, sketched out indeed by Ray and even by Linnæus himself, but elaborately developed by Jussieu, Decandolle, Lindley, and others.

The system at present adopted at Chelsea Garden is the Natural one, and the existing classification was devised and superintended by Dr. Lindley, who for some years, as will be afterwards noticed, occupied the position of Præfectus Horti and Professor of Botany to the Society of Apothecaries.

It appears that the Members of this Society, from the earliest periods of its history, have taken a very lively interest in the pursuit of botanical knowledge, for very shortly after their incorporation in the reign of James I., we find them engaging in periodical botanical excursions in the neighbourhood of London, and so early as the year 1632 the Society's "Herbarizings" had become an established annual custom. Thomas Johnson, the learned Editor of "Gerard's Herbal," and a member of the Society, has left a graphic account of several of these early excursions, proving the devotion to Botany of the "Socii itinerantes," as they were termed, and also showing the participation in that devotion by the Society at large.

The interest thus developed in the science of Botany among the members of the Society appears to have strengthened with time, and within half a century of their incorporation they succeeded in securing for themselves increased opportunities for the prosecution of their favourite study by acquiring a Botanic Garden of their own, namely, that which is situated on the northern bank of the Thames at Chelsea and forms the subject of the present Memoirs.

For more than two centuries the Society of Apothecaries have maintained this Garden at their own charge and for the sole object of the advancement of Botany as a science. From the name applied to it in the language of the period in the original conveyance by Sir Hans Sloane, of the "Physic Garden," it might perhaps be supposed that one, if not its chief purpose, was to grow plants to be converted into drugs for pharmaceutical use; but such an application of the contents of the Garden, although it was at one time partially contemplated, was afterwards expressly forbidden, and the word "Physic" was employed apparently in its general and etymological sense, as is indeed explained by Sir Hans Sloane himself in the document referred to, where he states that the Garden is to be maintained "for the manifestation of the power, wisdom, and glory of God in the works of creation."

It is true that one principal object of the Garden is to encourage the study of Botany on the part of medical practitioners and students, and the Society have always invited such visitors; and they have also, at various periods, appointed, for the purpose of instruction, Professors of Botany, and Demonstrators of the plants. It is also true that experiments have been and are still conducted for the purpose of verifying the genera or species of plants employed in medical practice, or of ascertaining the modifications or changes induced in medical plants by varieties of soil, temperature, and other similar conditions. But, in addition, the Garden has been always applied to general botanical purposes, and students of all Faculties have ready access to the advantages it affords for study, and, as will be seen in subsequent pages, ladies are now not only invited to visit and study in the Garden, but Prizes and Honorary Certificates are offered to such of them as are willing to enter into competition with members of their own sex for these honourable marks of distinction.

But while stating that the supply of pharmaceutical preparations forms no part of the design of the present Chelsea Garden, it is not improbable that the Society of Apothecaries may have previously possessed some other garden elsewhere for the sole object of growing medicinal herbs for the use of their laboratory which had been formed some years before. That there was a Botanic Garden at Westminster about the middle of the seventeenth century is made quite certain from some passages in the "Memoirs of the Life and Writings of John Evelyn, Esq.," the celebrated author of the "Sylva" and other works. In his Diary is the following statement: "1658, June 10th, I went to see the medical garden at Westminster, well stored with plants, under Morgan, a skilful botanist." It is extremely probable that this is the garden here alluded to, and that the Society of Apothecaries purchased the lease in order to obtain possession of the plants which it then contained, as it would thus certainly be a valuable acquisition for their new establishment at Chelsea.

It is much to be lamented that the early history of the Chelsea Botanic Garden is involved in great obscurity. The materials for forming such a history are extremely defective, but it is certain that the establishment of the Garden could not have been carried into effect without much previous deliberation and due investigation, by the appointment of committees, and by such other methods of proceeding as are usually adopted upon occasions of similar importance.

It would appear that the ground on which the Chelsea Garden is now situated was originally taken by the Apothecaries' Society as a spot on which to build a convenient barge-house for the ornamental barge which the Society (like the other City Companies) then possessed. In the minutes of the Society, bearing date June 26, 1673, it was ordered that a convenient barge should be constructed, and a spot chosen for a barge-house; and in July of the same year it was reported that no place could be found more convenient than that at Chelsea formerly proposed, and a lease of the ground was accordingly ordered to be taken. 1673

That this order was carried out in the same year appears from a recital contained in the release of the ground, granted many years after this period by Sir Hans Sloane, in which document it is stated that Charles Cheyne, Esq., by his indenture of lease, bearing date the 29th of August, 1673, did demise and grant unto the

Master, Wardens, and Society of the Art and Mystery of Apothecaries of the City of London, their successors and assigns, the piece and parcel of ground and premises therein mentioned, to hold from Michaelmas then next ensuing, unto the full end and term of sixty-one years, at the yearly rent of 5*l.*

1674 On the 21st of June of this year it is recorded in the minutes that several members, whose names are given in the succeeding paragraph, proposed to build a wall round Chelsea Garden at their own expense, with the assistance of such subscriptions as they might be able to procure; provided the Court of Assistants would agree to pay 2*l.* every year for ever to each of the six Herbarizings. This proposal was accepted, and the proprietors of the Laboratory Stock gave 50*l.* towards the building of this wall on the condition that they were to be allowed a piece of ground in the garden for herbs.

It is only right that the names of these liberal-minded Members of the Society should be recorded. They were as follow: Messrs. Sykes, Gardner, Rouse, Stratton, Reeves, Power, Warner, Watts, Ja. Rand, Leigh, Gaunt, Hollingworth, Lowry, and Hull.

Nothing more occurs relative to the garden until after the lapse of two years, when a circumstance is mentioned the nature of which is now difficult to understand. In the month of June of this year it was stated that the Court of Assistants agreed to take Mrs. Gape's lease of the garden at Westminster off her hands for the remaining two years for the sum of 16*l.* the rent being 2*l.* per annum; with the liberty of removing the plants to Chelsea garden. 1676

The name of the first gardener employed by the Society appears to have been Piggott, of whom however nothing more is known, but that his services were discontinued on the 16th of December; and it is recorded that Richard Pratt was chosen in his room, and was to be allowed lodging and a salary of 30*l.* per annum. As this must have been a handsome remuneration in those days, he was probably a man of considerable horticultural merit. In the autumn of this year directions were given that the garden was to be planted with the best fruit trees, and it is likewise stated that a good crop of herbs for the use of the Laboratory was furnished from the garden. 1677 1678

The management of this establishment began now to assume a more systematic form, a Committee being appointed for that purpose, consisting of twenty-one Assistants, thirty Liverymen, and twenty of the Yeomanry. It 1679 is difficult to conceive what could have been the motive for appointing so great a number of persons as are here mentioned to conduct such a concern, for it is sufficiently obvious that a numerous Committee is much more calculated to obstruct than to facilitate the despatch of business.

1680 In the month of January of this year, Mr. John Watts, a member of the Society, and one whose name has been mentioned as a contributor to the erection of the wall, was appointed to have the care and management of the Garden at 50*l.* per annum, besides the allowance of one or two labourers; and in the following year, a green-house or stove was erected in the Garden at an estimated expense of 138*l.* It was situated in the lower part of the ground, not far distant from the river.

1682 In the autumn of this year, Dr. Herman, Professor of Botany at Leyden, visited the Chelsea Garden, and proposed an exchange of plants, which proposal Mr. Watts went to Holland to 1683 carry into effect. About this time four cedar trees were planted in the garden near the river, being at the time of planting three feet high. Two of these remained healthy until lately; the two others, after continuing to live about a century, were cut down in consequence of their decayed state. One of the remaining trees died a few years since, and the only survivor is now (1878) unfortunately dying.

The expense of the garden, which is stated to have been 130*l.* annually, exclusive of the gardener's salary of 50*l.*, began now to be a matter of serious consideration, and in order to reduce this expense fresh proposals were made 1685 to Mr. Watts, the result of which was that he entered into articles of agreement (now in existence) with the Society, in which document he covenants to take upon himself the care, culture, and management of the garden, for the term of seven years from Michaelmas 1685; to repair and keep in repair the stove, greenhouse, and other buildings and utensils contained in it, and to make a catalogue of all the plants. For these services he was to be allowed by the Society the annual sum of 100*l.*, besides such expenses as he might incur in cultivating that part of the ground allotted to the use of the Laboratory Stock, with liberty to dispose of the fruit and supernumerary plants for his own benefit.

An order was made that the Master, Wardens, and Assistants should each be allowed to have a key of the Garden, at his own expense.

It is gratifying to observe the degree of

importance which this garden had attained in little more than ten years from its establishment, and its value appears to be recognised in some expressions used by Mr. Evelyn, who, in another part of the above-mentioned Diary, writes thus: "1685 August 7th I went to see Mr. Watts, keeper of the Apothecaries garden of simples at Chelsea, where there is a collection of innumerable varieties of that sort: particularly, besides many rare annuals, the tree bearing jesuit's bark, which had done such wonders in quartan agues. What was very ingenious was the subterraneous heat, conveyed by a stove under the conservatory, all vaulted with brick, so as he has the doores and windowes open in the hardest frosts, secluding only the snow."

1691 In a description of the gardens near London, given in December of this year and communicated to the Society of Antiquarians by the Rev. Dr. Hamilton, Vice-President, from an original MS. in his possession, and afterwards published in the 12th Volume of the Archæologia, the garden of the Apothecaries' Society is thus described: "Chelsea physick garden has great variety of plants both in and out of greenhouses: their perennial green hedges, and rows of different coloured herbs are very pretty; and so are the banks set with shades of herbs in the Irish stitch-way; but many plants in the Garden were not in so good order as might be expected, and as would have been answerable to other things in it. After I had been there, I learned that Mr. Watts, the keeper of it, was blamed for his neglect, and that he would be removed."

1693

Of the occurrences during the interval from 1685 to 1693 we have no information, but Mr. Watts's lease having now expired, the question was agitated in the Court of Assistants in the summer of this year, whether the Society should continue to support the Garden or not. It must be presumed that some material difficulty, probably as to the expected charges of it, presented itself to several of the Members. The good sense, however, of the major part of the Assistants prevailed, and the continuance of the Garden was determined upon, when Mr. Samuel Doody, a Member of the Society, undertook the care and expense of it at 100*l.* per annum.

1695

This arrangement, however, appears to have been of very short duration, for about two years afterwards, the state of the Garden having been taken into consideration, Mr. Doody was informed that the Court had reconsidered his proposals, and preferred giving him 100*l.* in lieu of his term, and of the agreement formerly made, instead of continuing the contract; but that they would grant a lease to him, or to any other Members who might desire it, for a term of 21 years. In consequence of this determination, a lease of the Garden for that term is stated to have been granted to Messrs. Doody, Petiver, Dare, and Bromwich, and at the same time it was directed that a pair of stairs should be made for those gentlemen, the intention of which direction, it may be supposed, was to facilitate communication with the river. This lease is not to be found among the archives of the Society, and it is probable, therefore, from this circumstance and from what followed in the succeeding year and subsequently, that its provisions were never carried out.

1696 The Minute Book of the Court of Assistants informs us that a treaty was now opened with Lord Cheyne relative to the Garden. As that nobleman had only a life-interest in the property, 1697 this treaty was not then completed; but in the ensuing year it must have been renewed, for the Court agreed to accept Lord Cheyne's offer to make their present term in the Garden sixty years for the sum of 75*l*. There were then about thirty-eight years unexpired, and this proposal must therefore have been intended as an addition of twenty-two years to the term. Notwithstanding the acceptance of this proposal by the Court as above stated, it is extremely doubtful whether such an agreement ever took effect. The original lease from Mr. Cheyne is not to be found, nor can any other writing upon the subject be discovered, and therefore whether this agreement was carried into effect by indorsement or by a separate deed cannot now be known. But what seems decisive upon the subject is, that Sir Hans Sloane's conveyance of the Garden, many years after, takes no notice of this extension of the lease, though the original lease of sixty-one years is distinctly recited. Had such an extension of the term then existed, it would have been too important a fact to be passed over in silence.

About this time a new proposal relating to the Garden was presented by Mr. Doody, and referred to a Committee, who reported that they approved the project, and agreed with him to carry it into effect. Of the nature of this proposal, however, nothing is now known.

1706

A period of several years now elapses without any intimation of the passing events respecting the Garden, which must have continued under the principal management of Mr. Doody until his death. This event took place some time in the year 1706. It may be presumed that Mr. Doody's skill and efficiency in the care of the Garden during all this time superseded the necessity of any particular interference on the part of the Society. As Mr. Doody was a man of considerable eminence both as a scientific medical practitioner and as a Botanist, a short biographical notice of him may not be unacceptable.

Mr. Samuel Doody, F.R.S.

He was a native of Staffordshire, but he settled in London as an Apothecary, and there is reason to believe that he acquired considerable practice. His botanical studies must have been principally confined to the examination of the plants growing in the vicinity of the metropolis, but his diligence in this pursuit was unexampled, particularly in the investigation of the flowerless or cryptogamous plants, his knowledge of which was superior to that of any other person of his day.

The early editions of Ray's Synopsis bear ample testimony to his labours. In the preface to that work (2nd Edit.) Mr. Ray notices him in the following terms, and the commendation of so distinguished a naturalist is a sufficient proof of Mr. Doody's great merit :—

"Samuel Doody, Pharmacopœius Londinensis, qui et opus ipsum plurimis speciebus ditavit; et alias, etiam tum species, tum observationes appendici reservavit. Non minus perspicax in plantis discernendis, quam industrius in indagandis; summis in hac Scientia viris æquiparandus."

Antoine Laurent de Jussieu, the celebrated Professor of Botany in Paris, and the founder of the Natural System of Botany which still bears his name, calls Mr. Doody "Inter Pharmacopœios Londinenses, sui temporis, Coryphæus."

In 1695 Mr. Doody became a Fellow of the Royal Society, which had been recently established (1662) in the reign of Charles II. There is a case of Hydrops Pectoris, written by him, in the Royal Society's Transactions for 1697; and some MSS. of his, on Medical and Botanical Subjects, are said to be preserved in the British Museum.

The mention of Mr. Doody's name, and this short tribute to his merits, afford a suitable occasion to observe that many of the Members of the Apothecaries' Society in the seventeenth and eighteenth centuries were distinguished for their scientific attainments, as will particularly appear in the course of the following pages. That all of them were good citizens and honourable men will be readily conceded from a consideration of the very constitution of the Society, and from a perusal of the Bye-laws by which the Society is governed, and there is every reason to believe that they were also efficient practitioners of the healing art. But it is not so generally known and acknowledged that some of them in their day held high positions in the scientific world, nor has sufficient commendation been bestowed upon the whole body, for their unselfish zeal in the promotion and encouragement of the study of Natural History, and especially Botany, as an essential element of medical education.

1707 In consequence of Mr. Doody's death, Messrs. Wyche, Andrews, and Petiver, were appointed to inspect the Garden, and a Committee was at the same time chosen to take into consideration its future disposal. This Committee made their Report in the month of March, but the document related solely to their treaty with Lord Cheyne, who it must be presumed had at this time an estate of inheritance in the Garden, for he offered to dispose of it to the Society for 400*l.*, which was eighty years' purchase of the ground-rent, but they refused to take it upon those terms. The Committee was then desired to consider whether it would be expedient to part with the Garden. In the month of June the Committee gave their opinion to the Court of Assistants that it was inexpedient to part with it, and it was ordered that the wharfing of it towards the river should be carried into effect as soon as it could conveniently be done.

Notwithstanding this determination to maintain the Garden, the expense of it must have been found too great for the Society to support in their corporate capacity, for in the following January a Common Hall of the Members was held for the purpose of proposing a subscription for the Garden, which was acceded to, and ninety persons gave in their names as subscribers. A Committee of the Court was appointed to meet the subscribers and consult upon terms, in consequence of which an agreement was entered into, and in the spring of this year a lease was granted to the subscribers, to which the common seal was affixed. The following are the conditions of this lease, which was made between the Society on the one part, and the subjoined Members on the other part, namely,— 1708

Robert Gower,
George Dare,
Joseph Nicholson,
Richard Springate,
Samuel Ryley,
Joseph Miller,
Robert Basket,
Benjamin Bouchier,
Henry Sherbell,
Richard Chapman,
John Channing,
John Perkins,
Thomas Robinson,
Robert Sheffield,
James Petiver,
Zachary Allen,
Isaac Garnier, jun.
Isaac Rand,
Derrick Barnevelt,
John Cox,

on behalf of themselves, and divers other Members of the Society; who had agreed to subscribe for seven years from Michaelmas 1707, so much as to make up the yearly sum of 100*l.* and upwards, such money to be employed in maintaining and improving the Physic Garden, for the benefit of the Society; and the above-mentioned subscribing Members were constituted Trustees for carrying the plan into effect. The Society therefore granted them a lease for seven years at an annual rent of 5*l.* upon condition that they repaired the buildings, and maintained, cultivated and supported the Garden in the best possible manner; reserving to the Master, Wardens, and Assistants, and also to every Member of the Society, who may be, or shall become subscribers to the Garden, full liberty of walking therein, either for recreation or instruction. Provision is also made for an annual audit by the subscribers of the accounts of the Trustees in order to secure a faithful application of the moneys subscribed.

1713 The plan of exonerating the Society from the charge of the Garden, by placing it in the hands of individual Members, does not appear to have answered any useful purpose; and the result was such as might have been expected. In the month of August of this year, the Trustees for the management of the Garden reported to the

Court of Assistants, that from the defect in the payment of subscriptions, and the great expense of supporting the Garden, they would not be able to continue its maintenance beyond their present term. The Court, considering that it would be for the honour of the Society and the benefit of the younger Members that the Garden should be carried on, ordered the following payments to be applied to the support of it; namely, that

	£	s.	d
Every Member should pay annually . . .	0	2	0
A Master on binding an Apprentice . . .	0	5	0
An Apprentice at the time of binding . . .	0	5	0
One Shilling in the Pound of the Laboratory Dividend.			
Every member dining at the private Herbarizing	0	1	0
Every fine payable to the Corporation to be Guineas instead of Pounds. The total of these was estimated at 90*l*. per Annum.			

1714 This year is memorable in the annals of the Garden, for affording the first intimation of a communication on its affairs with a gentleman, whose name and memory must always be held in high estimation by every lover of botanical knowledge, and especially by every friend to the scientific aspirations of the Society of Apothecaries. On the 1st of July a proposition was submitted to the Court by the Garden Committee, of waiting on Dr. Sloane (who had purchased the manor of Chelsea, of William, Lord

Cheyne in the year 1712) and inquiring his sentiments as to a future interest in the Garden. The immediate result of this conference does not appear, but it probably opened the way to that amicable and liberal settlement between the parties which took place a few years afterwards.

Sir Hans Sloane, M.D.

Dr. Sloane, afterwards Sir Hans Sloane, was a man of so much distinction in his day, and was so intimately concerned with the history of the Chelsea Garden, that a brief memoir of his life must be acceptable, not only to every member of the Society of Apothecaries but to the scientific world in general.

He was born in Ireland in 1660, but was of Scotch parentage, and while young his health was delicate, for he was troubled with spitting of blood from his sixteenth to his nineteenth year. As soon as his health permitted he came to London, and during four years devoted himself to the study of medicine and the collateral sciences. He was instructed in Chemistry by a disciple of Stahl, the celebrated inventor of the phlogistic hypothesis, and his fondness for Botany brought him the acquaintance of Ray, the distinguished naturalist. In 1683 he went to Paris, and while in that city he attended the Anatomical Lectures of Duverney and those on Botany by Tournefort. He afterwards spent a year at Montpellier, having been furnished with introductions by Tournefort to all the celebrated men of the University of that city, in the vicinity of which he passed much of his time in collecting plants, and, after travelling through Languedoc for the same purpose, returned to London in 1684.

He gave many of the plants and seeds he had collected to Ray, who described them in his "Historia Plantarum." In 1685 he was elected a Fellow of the Royal Society, and in 1687 a Fellow of the Royal College of Physicians of London. His curiosity, which had long been excited to see the wonderful productions of tropical countries, was somewhat unexpectedly gratified by his appointment of Physician to the Duke of Albemarle, who was going out as Governor of Jamaica, and he set out from England with that nobleman in 1687, arriving at Port Royal at the end of the same year. The Duke, however, unfortunately died soon after his arrival, and Dr. Sloane was therefore obliged to hasten his return to Europe, but nevertheless he made, during his residence of fifteen months in Jamaica, a very large collection of plants in that and the neighbouring islands. The plants which he brought home amounted to 800 species.

Dr. Sloane was appointed Physician to Christ's Hospital in 1694, and held that office for thirty years. In 1695 he married a lady of considerable wealth, the daughter of a London Alderman, and by her he had four children, two of whom died young, while two daughters survived their parents, and carried their wealth to the noble families of Stanley and Cadogan. The present names (Sloane, Cadogan, Stanley, &c.) attached to several streets, and squares, and estates, in and about Chelsea, sufficiently attest the magnitude and the value of the property thus derived from Sir Hans Sloane and his descendants.

In 1693 he was chosen Secretary to the Royal Society, in 1712 he was elected one of its Vice-Presidents, and in 1727 he succeeded the illustrious Sir Isaac Newton in the Presidency of the Society. In the last-named year he was appointed Physician to the King. In 1716, he had been created a baronet by George I. and appointed Physician-General of the forces; in 1719 he was elected President of the Royal College of Physicians, and he held that office till 1735. Thus loaded with well-deserved honours, and in the possession of abundant means, he at length retired, when eighty years old, to an estate he had purchased at Chelsea in 1720, and passed the remainder of his life in entertaining scientific men and in examining the treasures he had collected. He died in January, 1753, in the ninety-third year of his age.

Sir Hans Sloane directed that at his decease his Museum should be offered to the nation for 20,000*l.*, a sum which is said by him not to have amounted to a fourth part of its real value. At the time of his death, Sir Hans Sloane's cabinet contained 200 volumes of dried plants, and 30,600 specimens of other objects of Natural History, besides a library of 50,000 volumes, and 3566 manuscripts. He was, however, also known as a distinguished author, for he contributed many papers to the Philosophical Transactions, he assisted Ray in compiling his "Historia Plantarum Generalis," and published a large work in two volumes folio, on the "Natural History of Jamaica," in which he describes, and illustrates with many plates, the island itself, and its products, both animal and vegetable, and also alludes to its climate and the diseases of its inhabitants. He mentions in his preface that the whole undertaking had been submitted to Ray, and had met with his approval. A small Latin catalogue of the plants of Jamaica had been published by Sir Hans Sloane in 1696, and serves as a sort of Index to the large work. This work on the Natural History of

Jamaica must have involved great expense as well as literary and scientific labour. The plates of the plants, many of which had never been previously figured or described, are exceedingly well executed, and the arrangement and descriptions followed are those of Tournefort, with whom, as has been before mentioned, Sir Hans was personally well acquainted.

The lease of the ground, granted in 1708, being now near its expiration, a Committee was appointed in October to examine into the state of the Garden and to consider how it might be best managed. This Committee appears to have occupied much time in deliberation, which the numerous difficulties to be encountered must have rendered necessary. It was not until the following September, that, by an order of the Court of Assistants, it was determined that the Garden should be kept up for the present by the Corporation.

1714

In this year died Mr. James Petiver, a gentleman worthy of honourable mention in connexion with the history of the garden. The exact time of his birth is unknown. He was apprenticed to Mr. Feltham, Apothecary to St. Bartholomew's Hospital, and after his apprenticeship he settled in business in Aldersgate Street, where he continued for the remainder of his life. He was in large practice, and became

Mr. James Petiver, F.R.S.

Apothecary to the Charterhouse. He accumulated so large a collection of Natural History that, some time before his death, Sir Hans Sloane is said to have offered him 4000*l.* for it. After his death Sir Hans purchased it, and it came eventually to the British Museum. Mr. Petiver was a Fellow of the Royal Society, and among other labours he assisted Mr. Ray in arranging the second volume of his History of Plants, and that great naturalist gives this brief, but expressive and affectionate testimony to his merits. "Jacobus Petiver, non postremæ notæ botanicus, mei amicissimus."

Mr. Petiver must have been an extraordinary man, and it is almost incredible that, while busily occupied as a medical practitioner in the midst of London, he could contribute so largely as he did to the scientific literature of his day, and could accumulate such a multitude of specimens of all kinds in his museum. His writings, however, which are still extant, speak for themselves, as do likewise his collections which are in the British Museum. A genus of plants is named after him, viz. *Petiveria*, belonging to the Natural Order of Chenopodiaceæ.

His works on all branches of Natural History are very numerous, but they were all collected together in the lifetime of the author in two volumes bearing the following title:—

"Jacobi Petiveri Opera, Historiam Naturalem Spectantia; containing several thousand figures of Birds, Beasts, Fish, Reptiles, Insects, Shells, Corals, and Fossils; also of Trees, Shrubs, Herbs, Fruits, Fungus's, Mosses, Sea-weeds, &c., from all parts, adapted to Ray's History of Plants, on above Three Hundred Copper-plates with English and Latin Names. N.B.—Above One Hundred of these Plates were never published before. To which are now added Seventeen Curious Tracts, most of them so scarce as not to be purchased, which completes all he ever wrote upon Natural History."

The two volumes folio, of which this large work consists, contain an enormous amount of information, and the title-page states that the illustrations comprise no less than ten thousand articles engraved in the most accurate manner from originals which were the gifts of the most eminent persons in all nations. The letter-press contains a number of miscellaneous articles on different subjects, and essays on various matters connected with Natural History, some of which were contributed to the Royal Society. What Petiver calls his Gazophylacium[1] Naturæ et Artis, is a kind of *Catalogue Raisonné* of natural and

[1] Gazophylacium is a Greek word, or rather Greco-Persian word, Γαζοφυλάκιον, signifying a "depository of precious things."

artificial objects which he describes, dividing them all into ten *Decads*. His *Centuria Prima Musei Petiveriani* is a description of the objects contained in his Museum, the variety and extent of which may be estimated by his own definition, "Rariora Naturæ continens, viz. Animalia, Fossilia, Plantas, ex variis Mundi Plagis advecta, Ordine digesta, et Nominibus propriis signata." This large work was published in 1702, and must have been considered in its day a prodigy of industry, and it may be stated with truth of the illustrations that although they are of course deficient in the exquisite neatness and finish of the engravings of the present day, they are wonderfully well executed as well as being very accurate. Some of them indeed are marvels of elegance and accuracy.

Mr. Petiver officiated as Demonstrator of Plants to the Society of Apothecaries as early as the year 1709, but how long before that period he had held the appointment cannot now be ascertained. He probably resigned it to, or at least was assisted in that office by, Mr. Rand, a member of the Society, and also a distinguished botanist. Mr. Petiver died on the 20th of April, 1718, Sir Hans Sloane and other eminent men, in token of their esteem, attending his funeral as pall-bearers.

1718

The new ordinances of the Corporation containing some provision for the support of the Garden, the Master, Wardens, and three other members, were deputed to wait on Sir Hans Sloane, to acquaint him with the details. It must have been upon this occasion that the following statement, copied from a written paper among the Sloane MSS. in the British Museum, was sent to Sir Hans Sloane, probably for his approbation. As it differs materially from the estimate before mentioned, and is several years subsequent to it in date, it will be important to insert it. It is entitled, "a note of paragraphs forty-six and forty-seven, with a calculation of the produce, communibus annis."

To apply the following Fines to maintain the Garden at Chelsea, and the private Herbarizings.

	PRODUCE. £	s.	d.
5*s*. on taking an Apprentice . . . 7*s*. 6*d*. on becoming an Apprentice .	30	0	0
10*s*. on being made free	8	0	0
20*s*. on admission as Assistant and Liveryman . . . . . . 40*s*. on a Foreigner admitted by Redemption	20	0	0
A moiety of Quarterage from Members .	35	0	0
1*s*. in the Pound of Laboratory Dividends, but this not to exceed 20*l*. in the whole	15	0	0
Rent from Tallow-chandlers' Company for Barge-house, from Mr. Abbot, painter, Chelsea	4	10	0
	£112	10	0

This interview with Sir Hans Sloane appears to have been attended with the best effects. When men of candour and public spirit meet together to promote designs of science and general utility the difficulties and obstacles, which readily present themselves to worldly-minded characters, soon vanish from view, and are never allowed to obstruct the progress of honourable and disinterested negotiations. Such was the case in the instance now referred to, and within the short space of a few weeks the Court of Assistants were informed that Sir Hans Sloane was ready to settle the Garden upon the Society on the terms proposed.

About this time it was ordered that a ticket of admission to the Garden should be given to each of the proprietors of the Laboratory Stock in consideration of their contributing 15*l.* per annum from that stock towards the expense of the Garden. This sum of 15*l.*, corresponding with the statement given in the Sloane MS., is evidently the same as that referred to in the latter document.

1722

The deed of conveyance of the Garden at Chelsea from Sir Hans Sloane was laid before a Court of Assistants on the 8th February, 1722, approved by them, and ordered to be sealed.

The most important covenants contained in this conveyance are the following, namely,—

The release is dated on the 20th of February, 1722, and is made between the Honourable Sir Hans Sloane, Baronet, President of the Royal College of Physicians on the one part, and the Master, Wardens, and Society of the art and mystery of Apothecaries of the City of London, on the other part. It recites the original lease from Lord Cheyne, and also the great expense which the Society had incurred, in furnishing and carrying on the Garden, as a Physic Garden, ever since that lease was granted. It states that the Fee and Inheritance of the ground and premises were then vested in Sir Hans Sloane and his heirs. It further declares, that to the end the said Garden may at all times hereafter be continued as a Physic Garden, and for the better encouraging and enabling the said Society to support the charge thereof, for the manifestation of the power, wisdom, and glory of God in the works of the creation, and that their apprentices and others may better distinguish good and useful plants from those that bear resemblance to them, and yet are hurtful, and other the like good purposes ; the said Sir Hans Sloane grants, releases and confirms unto the said Master, Wardens, and Society, and their successors, all that piece or parcel of arable and pasture ground, situate at Chelsea in the county of Middlesex, at that time in their possession, containing three acres, one rood, and thirty-five perches, with the green-house, stoves, barge-houses, and other erections thereon, to have and to hold the same for ever, paying to Sir Hans Sloane, his heirs and assigns, the yearly rent of 5*l.* and rendering yearly to the President, Council, and Fellows of the Royal Society of London, fifty specimens of distinct plants, well dried and preserved, which grew in their garden the same year, with their names or reputed names; and those presented in each year to be specifically different from (those of) every former year until the number of two thousand shall have been delivered. It is further provided, that if these conditions be not fulfilled, or if the Society shall at any time convert the Garden into buildings for habitations, or to any other uses, save such as are necessary for a Physic Garden, for the culture, planting and preserving of trees, plants, and flowers, and such-like purposes: then it shall be lawful for Sir Hans Sloane, his heirs and assigns, to enter upon the premises, and to hold the same for the use and benefit, and in trust for the said President, Council, and Fellows of the Royal Society, subject to the same rent, and to the delivery of specimens of plants, as above mentioned, to the President of the College or Commonalty or Faculty of Physic in London; and in case the Royal Society shall refuse to comply with these conditions then in trust for the President and College of Physicians of London, subject to the same conditions as the Society of Apothecaries were originally charged with.

Power is also reserved for the President, or Vice-President of the Royal Society, and for the President, or Vice-President of the Royal College of Physicians, once or oftener in every year, to visit the said Garden, and examine if the conditions above specified are duly observed and complied with.

1722 In the spring of this year, a Committee consisting of the Master, Wardens, and nine other Members of the Court of Assistants, who had been appointed for the purpose of putting the Garden in order and taking upon themselves its management, reported to the Court their opinion that the present gardener should be dismissed, and upon their recommendation Mr. Philip Miller was appointed in his room. The name of the person discharged is not mentioned, but their choice of a successor was peculiarly happy, and reflected great credit upon the discernment of the Committee. The biographers of this great botanist and horticulturist, Philip Miller, say that he succeeded his father in the office, but it does not appear what authority they have for this assertion. The records of the Society are silent upon the subject, and the statement is therefore probably a mistake.

Regulations for conducting the affairs of the Garden were drawn up by this Committee, and approved by the Court on the 21st of August. These regulations are entered at length in the Minute Book of the Garden Committee, dated the 26th October.

In the month of August of this year, the first presentation was made of fifty plants to the Royal Society, agreeably to one of the covenants contained in Sir Hans Sloane's deed of conveyance.

The Society having now obtained complete possession of the Garden and a permanent interest in its maintenance, promptly turned their attention to its state and condition, which it may readily be supposed must have now become, both as to repair and cultivation, very unsatisfactory. It had been for many years before this time in the hands and under the management of private individuals, who could not be expected to expend more upon it than was absolutely necessary; and, when the Society again resumed the control of it, the uncertainty of their tenure at that time would have deterred them from any material expenditure.

To facilitate the means of repairing the Garden and putting it in order, recourse was again had to a Common Hall, or meeting of the Members at large. This was held on the 27th of June, when it was agreed that every Freeman should pay 2*s.* 6*d.* per quarter, 2*s.* of which was to be applied to the use of the Garden; but as this payment was afterwards thought too large, it was by a subsequent Common Hall (16th January, 1723) reduced to 6*s.* per annum; and in order to complete the necessary sum, 40*l.* was to be allowed annually by the Society from their corporate funds. Soon after this time the Royal College of Physicians of London very liberally presented the Society with 100*l.* for the use of the Garden. Though the amount of the sums raised by these means could not be very considerable as an immediate supply, yet it seems to have answered the desired purpose, as we meet with no complaint of pecuniary difficulties for some years afterwards.

The Garden Committee now recommended that some person should be appointed in the character of Præfectus Horti, or Director of the Garden, whose duty it should be to visit it often, and to undertake its superintendence and inspection. Mr. Isaac Rand, a Member of the Society, was appointed for that purpose; an eminently capable man, and one who had for several years been a very zealous cultivator of botanical science. By way of recompense for his services he was to be admitted a Member of the Court of Assistants without a fine, to be excused from all other offices, and to have a salary of 50*l.* As he was at the same time Demonstrator of Plants, this salary, it may be presumed, included the income derived from both offices, and to this sum was afterwards added 8*l.* per annum, being the interest of a legacy of 200*l.* bequeathed by Mr. John Meres, Clerk to the Society, for the benefit of the Demonstrator for the time being. The latter was to attend in the Garden during the six summer months, at least twice in each month, to demonstrate the plants to such persons as should then attend, and to execute such other matters as were enjoined by former orders.

1724

Mr. Isaac Rand, F.R.S.

From a Report of the Garden Committee, it

1727

appears that the whole expense of the Garden for this year was 172*l.* 13*s.* 7*d.*

1728 Directions had been given, so far back as the year 1707, for carrying into effect the wharfing of the Garden towards the river. Either this was not done at that time, or it was executed so imperfectly that the structure must have fallen into decay; for we find an order now given that a wharf should be built along the river before Chelsea Garden, and that 1000*l.* be borrowed for that purpose on the Society's Bonds, from such of the Members as might be willing to contribute to the outlay.

1730 In this year appeared the first published catalogue of the officinal plants in the Chelsea Garden. It was drawn up by Mr. Philip Miller, the Gardener, and is entitled "Catalogus Plantarum Officinalium, quæ in Horto Botanico Chelseyano aluntur." It is a small book, and is written in Latin, but every plant mentioned has an English name appended to it, so that this catalogue must have been a very useful manual for those who wished to make themselves acquainted with the contents of the Garden and with plants in general. The division of the plants adopted by Mr. Miller is very simple, only two sections being admitted, namely, the Herbs and Undershrubs (*Herbœ et Suffrutices*), and the Trees and Shrubs (*Arbores et Frutices*). This crude arrangement is derived from Ray. Each section is classified alphabetically, the first section containing the names of 405 plants, and the second of 94. The plants are not described in detail, but references are given in each case to the authorities where descriptions or figures may be found, as, for instance, to Tournefort, Ray, Bauhinus, Boerhaave, Fuchsius, Lobel, Martyn, Gerarde, &c.; and the numbers accompanying each name in the Catalogue probably refer to corresponding numbers affixed to the plants in the Garden. The plants themselves are of a very miscellaneous character, but they comprise a great number of those which were employed at the time, and are still employed in medicine, or were supposed to possess some therapeutical virtue, together with ornamental herbs, shrubs, and trees, cultivated for the beauty of their flowers, the utility of their fruit, or other similar claims to the notice of botanists.

1732

Among the acting Members of the Committee for conducting the affairs of the Garden there now occurs the name of James Sherard, a Botanist of eminence, who withdrew from the Society about this time, in consequence of having obtained a diploma of Doctor of Medicine, and practising as a Physician. Dr. James Sherard was so distinguished as a cultivator of science that no apology is necessary for the introduction of the following memoir of his life and that of his brother.

Dr. James Sherard, F.R.S.

James Sherard was the son of George Sherard of Bushby, in the county of Leicester, and was born probably about the year 1666, as it appears from the Court Books of the Society that he was bound apprentice to Mr. Charles Watts, on the 7th February, 1682. Mr. Watts, although in practice as an Apothecary, had been, about a year before the date referred to, appointed to the care and management of the Garden at Chelsea, which circumstance must have given his apprentice great opportunities for cultivating a taste for botanical pursuits, and no doubt laid the foundation of his future eminence. Mr. Sherard practised for some years as an Apothecary in Mark Lane, London, where he occasionally made a public exhibition of rare plants, in the study and cultivation of which he was a great proficient. He acquired considerable wealth by his practice, and in the latter part of his life, having taken the degree of M.D., and having been elected a Fellow of the Royal Society, and of the Royal College of Physicians, he retired to Eltham in Kent, where he continued to pursue his favourite amusement, the cultivation of valuable and uncommon plants. The garden belonging to him in that locality became celebrated by the Catalogue of its contents, drawn up and arranged by Dillenius, Professor of Botany at Oxford, and published under the title of "Hortus Elthamensis, sive Plantarum Rariorum, quas in horto suo Elthami in Cantio collegit vir ornatissimus et præstantissimus Jac. Sherard, M.D., Soc. Reg. et Coll. Med. Lond. Soc., Guilielmi P.M. frater. Delineationes et descriptiones, quarum historia vel planè non, vel imperfectè, rei herbariæ scriptoribus tradita fuit. Auctore Jacobo Dillenio M.D. (London, 1732)." (The Garden of Eltham, or of the rarer plants which the most distinguished and excellent man, James Sherard, M.D., Fellow of the Royal Society and of the College of Physicians of London, brother of William (Sherard) of pious memory, collected in his garden at Eltham in Kent. Engravings and descriptions, the history of which has been altogether omitted or imperfectly described by writers on Botany. Edited by James Dillenius.)

This large and expensive book is in two folio volumes, and is illustrated by 324 plates, which, though uncoloured, are exceedingly well executed, and prove that the Eltham garden was worthy of the admiration bestowed upon it. The descriptions of the plants, which are arranged alphabetically, are written in Latin. In a letter to Sir Hans Sloane, dated December, 1732, and published in the "Gentleman's Magazine," Dr. Sherard says "I send herewith a copy of the Hortus Elthamensis which Dr. Dillenius is now publishing. You will see that he has not studied to adorn either his book or my garden, his chief care having been to improve and advance the knowledge of Botany."

Although Dr. James Sherard did not appear himself as an author, he was an excellent scientific botanist, and he gave his liberal assistance towards the diffusion of knowledge by assisting the efforts of others. It is said that a Natural History of Carolina, by an author named Catesby, was brought out by Dr. James Sherard's pecuniary and scientific help; and the costly Hortus Elthamensis was published by Dillenius under the same circumstances of friendly co-operation.

He died on the 12th of February, 1737, and was buried at Evington in Leicestershire; and the following inscription appears on his monument in Evington Church:—

M.S.
JACOBI SHERARD, M.D.
Colleg. Medic. Lond. et Soc. Reg. Soc.
viri multifariâ doctrinâ cultissimi
in rerum naturalium Botanices imprimis
scientiâ
penè singularis
et ne quid ad oblectandos amicos deesset
artis Musicæ peritissimi.
Acceperunt illi in laudis cumulum
mores Christiani, vitæ integritas,
et erga omnes comitas et benevolentia.
Obiit pridie id. Feb., A.D. MDCCXXXVII
annos natus LXXII.
Uxor Susanna, Richardi Lockwood, Arm. filia
optimo marito
hoc monumentum mœstissima posuit
et sibi quæ ob. 27th Nov. 1741, ætat. 72.
et juxta maritum sepulta est.

The name of James Sherard often occurs in the Synopsis of Ray, and in the Preface to the third edition of that work the author gives the following eulogy of Sherard in conjunction with another botanical friend. "Præcipuas verò, in hoc opere, partes occupant duumviri D. Richardson et D. Jac. Sherardus, amici nostri honoratissimi, qui, de industriâ, crebris institutis itineribus botanicis, Plantarum Angliæ familiam plurimùm auxerunt, plantas dubias, earumque loca minus certa restituerunt, et species demùm novas nondum descriptas, ipsi invenerunt."

In a MS. book (Sloane MSS. 3347) entitled Adversaria, by Mr. James Petiver, to whom reference has already been made (page 30), is an entertaining description of a botanical excursion made by James Sherard and himself into Kent. These two distinguished botanists travelled in a chaise with two horses. At Winchelsea they were regaled at the Mayor's house, and, the place not affording any wine, they had several bowls of excellent punch made by the Mayoress, "each bowl of which," the account says, "was better than the former one."

James Sherard's name is perpetuated to botanists in a genus of plants called *Sherardia*, of the Natural Order of Rubiaceæ, a species of which, *S. arvensis*, is a common weed in our corn-fields, "a little insignificant weed," as it is designated in *Loudon's Encyclopædia of Plants*, "by no means worthy to be consecrated to the memory of so celebrated a man."

James Sherard was the younger brother of William Sherard, D.C.L., also a distinguished botanist, and the two names are often confounded together. William Sherard, however, was born in 1659, and died in 1728, and was buried at Eltham. He was a Fellow of St. John's College, Oxford, but he devoted great attention to botanical pursuits both at home and abroad, and in Sir Hans Sloane's Catalogue of Plants (MS.) there is a long list of seeds sent by him to Sir Hans; and in another MS. are several of William Sherard's letters to Sir Hans from Ireland, Leyden, the Hague, Venice, Rome, and Paris, chiefly on botanical subjects, and several both on Botany and Greek literature from Smyrna, where Sherard was Consul from 1704 to 1715. By his last will he left an endowment for the Professorship of Botany at Oxford, and appointed Dillenius the next Botany Professor, and he also left all his books on Botany and Natural History, together with all his drawings, paintings, and dried plants, (particularly his Herbarium and Pinax) to be deposited in the Library of the Physic Garden. The Sherardian Professorship of Botany (for so it is still called) and the beautiful Botanic Garden, now in the most flourishing condition, belonging to the University of Oxford, constitute worthy and lasting memorials of the scientific acquirements as well as of the liberality of Dr. William Sherard.

William Sherard and James Sherard were so intimately associated together, both by their close relationship and by the similarity of their scientific pursuits, that it is impossible to mention the life and labours of one without referring to the other. William was so distinguished a botanist as to have acquired the name of MAGNUS; and Linnæus, who visited the Oxford Garden in 1736, and inspected its contents together with the herbarium and library left by Sherard, alludes in the following eulogistic terms to the two distinguished brothers:—"Consul Gulielmus Sherardus, agnomine apud botanicos MAGNUS dum suam vitam, se ipsum, et omnia sua rei herbariæ consecravit, immortalem apud botanicos obtinuit gloriam, quæ perennabit virens et florens dum vivunt et florent plantæ, dato præsertim a J. J. Dillenio Phytopinace Sherardiano, uti quoque quondam Hortus Elthamensis Sherardi magni et Johannis[2] fratrum, Dillenio authore, sine pari prodiit." (Consul William Sherard, called the GREAT among botanists since he devoted his life, himself, and all his possessions to botanical pursuits, has obtained among botanists immortal glory, which will continue bright and flourishing as long as plants live and flourish, especially since the publication by J. J. Dillenius of Sherard's Catalogue of Plants, as also, in former years, the Hortus Elthamensis of the brothers, the great Sherard and John,[2] was without a rival.)

---

[2] It should be James (Jacobi).

In reference to the Latin quotations (to which translations are in most instances appended) so often introduced in the present Memoirs of the Chelsea Garden it is perhaps necessary to explain that the Roman language was the ordinary medium of communication between scientific and literary men for many centuries after the revival of learning, and not only were scientific books written in that tongue, but it formed the means of conversation and correspondence between the learned of different countries. Whether the disuse of Latin for this purpose in the present day is an advantage, may fairly be a matter of doubt, for it is now necessary for the scholar to learn a great number of modern European languages where a single ancient one formerly sufficed. Linnæus himself, whose acquirements in the whole range of science were no less than gigantic, understood only two languages, namely, Latin and his own, the Swedish, and, as the latter is but little spoken or read out of Sweden, all his letters to foreigners and most of his books were written in Latin. Although he visited many of the capitals and the countries of Europe, including England, France, Lapland, and Holland, he learned the language of none of them, and he knew nothing even of

Dutch, although he spent three years in Holland superintending a large garden and engaging in scientific studies with other distinguished men such as Boerhaave, Gronovius, Van Swieten, Lieberkühn, Albinus, &c. It is therefore not a matter of astonishment that the early members of the Apothecaries' Society were in general well acquainted with Latin, and that the more eminent among them wrote and spoke in that language. It is only an act of justice moreover to observe that the Society have consistently encouraged and even enforced the study of the Latin language upon their alumni. While the Society's legal jurisdiction extended only to the education of their own members, a knowledge of Latin was indispensable in the case of candidates for examination; and since the existence of the Court of Examiners, under the Act of 1815, not only has Latin been a compulsory subject of examination for the Licence of the Society, but the present Examination in Arts, now rendered imperative on all medical students, originated in Regulations made by the Court of Examiners many years ago.

1732 We have now arrived at a period of great importance in the history of Chelsea Garden. We are totally ignorant as to the state of the buildings belonging to the garden before this period, whether in reference to the cultivation and preservation of the plants, or to the comfort and accommodation of the Gardener. But the Society at this time formed a design, and began to carry into effect a plan for erecting a fabric calculated to serve both these purposes, upon a scale of such extent and elegance as should be calculated to reflect honour upon their members, and should also embrace the most approved methods then known of raising and bringing to maturity the more delicate productions of foreign countries, by imitating as nearly as possible the various temperatures enjoyed by such plants in their native climates.

As the outlay attending this extensive building seemed likely to be much greater than the Society's funds could afford, a subscription among the members was again resorted to, which amounted to 549*l.* 14*s.*, the Corporation agreeing to contribute a sum not exceeding 500*l.* for this important purpose.

A Court of Assistants, held on the 24th of August, approved the agreement entered into by the Garden Committee for building the green-house and two hot-houses for the sum of 1550*l.* but the addition of some flues, proposed by Mr. Miller, the Gardener, caused an increased charge of 125*l.* 18*s.*, which sum was further augmented by the erection and completion of two staircases, the expense of which was estimated at 140*l.* 18*s.* Thus the total expense, as far as can now be ascertained, including the sum of 76*l.* 5*s.* paid to the Surveyors, was 1891*l.* 16*s.*

The buildings were immediately undertaken, and were completed in less than two years.

In the bed of a Plinth to the middle break in the front of the present green-house lies, inclosed in lead, a copper plate, on which is engraven the following inscription, viz. :—

> This was the first stone laid by the hands of Sir Hans Sloane, Donor of this Garden to perpetuity, for the improvement of useful knowledge.
>
> *August* 12*th*, 1732.
>
> MR. WILLIAM WITHERS,
> *Master of the Society of Apothecaries.*
>
> MR. RALPH FORSTER, } *Wardens.*
> ROBT. HARRIS, ESQ. }
>
> EDWARD OAKLEY, *Architect.*

During the present year the Court of Assistants made an order which still further evinced their zeal in forwarding the cultivation of botanical knowledge. It was to the effect that 20*l.* per annum be paid by the Society towards the expense of sending a person to Georgia, to collect trees and plants, and to make experiments concerning the best means of raising them in England; and this sum was in the following year ordered to be paid to the Trustees for that colony.

1733 As a monument of gratitude to the munificence of Sir Hans Sloane, the Court of Assistants (August 23rd) came to a determination to cause a statue of him to be erected in marble, which was originally placed in the front of the green-house. Its position, however, was afterwards changed, and in 1748 it was removed and fixed on a pedestal in the garden, nearly in the centre, facing the principal building. The cause of this change seems to have been a doubt as to the strength of the front wall being equal to the weight of the statue, for it appears that the opinions of three surveyors were required upon this question. The statue was the work of the celebrated Michael Rysbrach, and was finished in the year 1737, at the cost of 280*l.*

1736 Mr. Zachary Allen, a Member of the Society, and one of the lessees of the Garden in 1707, bequeathed 50*l.* for the service of the Garden.

In this year Linnæus visited the Chelsea Garden, and the circumstance is thus mentioned in his Diary :—" Miller of Chelsea permitted me to collect many plants in the Garden, and gave me several dried specimens collected in South America." This event is interesting, as showing the high estimation in which the Chelsea Garden was held by the scientific men of foreign countries as well as our own, and its value may be further appreciated by the fact that only two Botanic Gardens in England were mentioned or visited by Linnæus, namely, that at Oxford, then under the care of Professor Dillenius, and that at Chelsea, under the management of Rand and Philip Miller, to both of whom, together with Hudson, also a Demonstrator at Chelsea Garden, Linnæus often refers in the most commendatory terms.

As the researches and the writings of Linnæus soon effected a scientific revolution, not only in the Chelsea Garden, but in all the gardens of the civilised and educated world, and indeed in every region of the habitable globe, it would be unpardonable not to pause for a brief reference to the career of this distinguished man, who might ask with Æneas, when recalling to mind his manifold adventures,—

> "Quæ regio in terris nostri non plena laboris?"

and the remarks about to be made may not be amiss in the present day when perhaps too little respect is paid to the memory of the Swedish naturalist.

Linnæus.

Carl Linnæus, or Carl von Linné, as he was afterwards called when he was ennobled by the King of Sweden, rose by his own prodigious efforts to the very highest position in the scientific world. Being educated as a physician, and for some time practising that profession, he wrote largely on Medicine, and endeavoured to introduce the same principles of systematic arrangement into Pathology and Therapeutics as those he had so successfully applied to the study of Natural History. But it was in the latter department that he particularly shone, and, as we have just intimated, there is not a botanical or zoological garden, nor a collection of botanical, mineralogical, or zoological specimens in any city or town in the world wherein the traces of his marvellous genius may not be discovered. It is not asserted, of course, that all his conclusions were correct, or that any one of the numerous principles of classification he proposed was faultless, but it may be safely and truly said that he illuminated the whole domain of Natural History with new and striking observations and discoveries, and arranged into harmony and order a multitude of details which lay previously heaped together in chaotic confusion.

It is not consistent with the plan of these Memoirs to enter at large into the merits of Linnæus and his multifarious labours; and, even in reference to the study of Botany, the remarks made must be brief. It will suffice therefore at present to state that his knowledge of plants was most varied and extensive, that his labour in acquiring that knowledge was indefatigable and incessant, and that he seemed to grasp intuitively the principles on which vegetable forms are constructed and the laws on which their classification is based. In making these remarks, it is not intended to institute a comparison between his System (the Linnæan) and that which is called the Natural System, of which something will be said hereafter. There cannot, however, be the slightest doubt that at the period which our history of the Chelsea Garden has now reached (namely the middle of the eighteenth century) the science of Botany was in its infancy, if indeed it could be said to exist, and that it was only after the publication of Linnæus's works that it began to assume the importance which it claims in the present day as one of the most delightful and engaging of intellectual pursuits.

Alluding, as briefly as possible, to the novelties introduced by Linnæus into the study of plants, it may be stated that his classification was founded, primarily, upon the presence or absence of the reproductive organs, namely the stamens and pistils, the true functions of which, before his time, were very imperfectly understood, if they were understood at all. Having then separated the plants having no stamens or pistils (or as he called them the *Cryptogamic Plants* [3]) from those which possessed both (the *Phænogamic Plants* [3]), he divided the latter into CLASSES, founded upon the number, relative size, situation, and other peculiarities, of the Stamens; and these classes again were subdivided into ORDERS, founded upon the number of the Pistils, and upon other peculiarities of the reproductive organs. The Classes were only Twenty-four in number, and thus the Linnæan System possessed a merit which has never been denied to it, namely that of simplicity, and its principles could readily be learned almost by the merest tyro.

But the invention and promulgation of the Linnæan, or Sexual System of Plants, was only

[3] Κρύπτω, I conceal, γάμος, a marriage; φαίνω, I appear, γάμος, a marriage. Linnæus, whose ideas were always poetical, compared the union of stamens and pistils to a marriage; and where no such organs were to be found he considered the marriage to be *concealed*, though he did not deny its existence.

one out of many services rendered to Botany by the illustrious Professor of Upsala, for he revolutionised and simplified the previously heterogeneous nomenclature of plants; gave to all of them definite names, assigning to each plant a genus and a species; threw into the same genus many forms which had before been distributed among several genera, and, *vice versâ*, split up several single genera into different genera and species; reduced varieties to species, and pointed out the true limits of varieties, species, and genera; defined the characters by which genera might be distinguished from one another; and in fine, conferred inestimable services upon the study of every department connected with the vegetable kingdom. The student of the present day, whenever he names in scientific language any tree or flower, and, it may be added, whenever he names in a similar way any member of the animal kingdom, consciously or unconsciously employs the language of Linnæus.

1737 The names of the subscribers to the building of the green-house were ordered to be painted on a suitable frame, and hung up in some conspicuous part of that edifice.

1739 In this year Mr. Isaac Rand, the Præfectus

Horti and Demonstrator in Botany, presented to the Court of Assistants his Catalogue of the Officinal Plants in the Garden, entitled " Horti Medici Chelseiani Index Compendiarius, exhibens Nomina Plantarum quas ad Rei Herbariæ præcipue Materiæ Medicæ scientiam promovendam ali curavit Societas Pharmacopœiorum Londinensium. Conscripsit Isaacus Rand, Pharmacopœius Londinensis, Regiæ Societatis Socius, Horti Præfectus et Prælector Botanicus." (A compendious Index of the Chelsea Physic Garden, exhibiting the Names of the Plants, which the Society of Apothecaries of London have directed to be cultivated for the promotion of the science of Botany and especially of Materia Medica. By Isaac Rand, &c.)

It would appear from some passages in the biography of Mr. Philip Miller, that Mr. Rand prepared this Index in consequence of his feeling hurt at the publication of the Catalogue of the contents of the Garden by the former, who, as Mr. Rand considered, had encroached upon his province, he being the Præfectus Horti and Botanical Demonstrator, while Mr. Miller was the Gardener. Both Mr. Miller and Mr. Rand, however, were excellent Botanists.

Rand's Index is a larger book than Philip Miller's Catalogue, and consists of 214 pages, while the Catalogue has only 152. The Index is written entirely in Latin, and does not give the English names of any of the plants, which omission must have diminished the value of the book very much in the eyes of the apprentices and the beginners in Botany, for whose use it is said to have been principally designed. It is very laboriously compiled, and gives a multitude of synonyms of all the plants. Ray and Tournefort are the principal authorities referred to, but many others are also quoted. The arrangement is alphabetical, and no scientific classification is adopted, even into Herbs and Trees, as is the case in Philip Miller's Catalogue. Although the title indicates that the chief plants described are those employed in Medicine, yet a very large proportion of them have no claim whatever, at any rate in the present day, to the possession of any therapeutical powers. Many of such plants, however, though now regarded as inert, were once supposed to be useful as medicinal agents, and it must be admitted that a few, which were formerly regarded as useless, have been found to possess active principles which the researches of modern chemistry have discovered and developed into valuable drugs.

Mr. Rand was a Fellow of the Royal Society, and he enjoyed the friendship and esteem of

Linnæus. A genus of plants was named after him *Randia*, belonging to the Natural order of Rubiaceæ, and found in tropical countries.

1739

Mr. Samuel Dale, a Member of the Society, author of a Pharmacologia, and some other works, left to the Society by his will a legacy of books which were ordered to be kept in proper presses to be provided by the Committee, and the clause in Mr. Dale's will relating to this bequest was ordered to be entered in their minutes.

A short biographical memoir of this gentleman whose professional and literary abilities would have rendered him an ornament to any sphere of life, must be regarded only as a just tribute of gratitude to one who was both a Member and benefactor of the Society of Apothecaries.

Dr. Samuel Dale.

Samuel Dale was born about the year 1658, at Braintree in Essex, in which place he practised for some years as an apothecary. In 1693 he first published his Pharmacologia, in an octavo volume. This work was afterwards republished in quarto, and went through several editions both in England and in foreign countries. His first edition was among the earliest scientific books written on the subject of drugs and medicinal preparations, and it specially displayed great botanical knowledge. The subsequent editions were much improved by the author, so as to render the book, not only at that time, but even at present, a work of considerable value.

The third edition, published in 1737, bears the following title :—

Samuelis Dalei M.L., Pharmacologia, seu Manuductio ad Materiam Medicam; in qua Medicamenta Officinalia Simplicia, hoc est Mineralia, Vegetabilia, Animalia eorumque partes in Medicinæ Officinis usitata, in Methodum Naturalem digesta succincte et accurate describuntur. 1737.

(The Pharmacologia of Samuel Dale, Licentiate of Medicine, or Guide to the Materia Medica; in which the simple Officinal Medicines, namely Minerals, Vegetables, and Animals, and the parts of them used in Medical establishments, brought together into a Natural Method, are succinctly and accurately described.)

In this edition the author states that his first edition appeared forty years previously. In the botanical part, which forms a very large part of the volume, Dr. Dale offers his thanks to many distinguished botanists for their assistance and co-operation, and among others, to Sir Hans Sloane, John Ray, William Sherard, Samuel Doody, James Petiver. The work is written in Latin, but the English names are given to each substance. The System of Tournefort is adopted, as might be expected, for the third edition of Dale's Pharmacologia was published in 1737, and the Linnæan System had not then been generally promulgated. Dr. Dale's Pharmacologia displays a great amount of learning and research, but much of the contents would be considered in the present day as involving a waste of labour, inasmuch as many of the substances described are now regarded as inert. But perhaps our own Systems of Pharmacology are not so thoroughly weeded of useless or superfluous matters as to justify the present generation in ridiculing the polypharmacy of their predecessors.

Mr. Dale also republished a work of Silas Taylor, entitled, "Antiquities of Harwich and Dovercourt," which is valuable for its very accurate notices on Natural History.

In 1730 Mr. Dale obtained the degree of Doctor in Medicine, became a Licentiate of the Royal College of Physicians, and removed to Bocking in Essex, where he practised until his death, on the 6th of June, 1739, in the eightieth year of his age. There is a portrait of him in the Court-room at Apothecaries' Hall. He was buried in the Dissenters' burial-ground at Bocking.

The conditions stipulated by Dr. Dale, in relation to his devise of books and dried plants, were as follow, "That the Master and Wardens shall, within twelve months next after his decease, make or erect proper conveniences in their Physick Garden at Chelsea, for the reception thereof, and under such regulations for the keeping and preserving them, as shall be agreed on and approved by Sir Hans Sloane and his executors after named." Sir Hans Sloane having been consulted on this occasion, his opinion was that it would be most proper to have a press or presses made sufficient to contain the whole of the books and plants, in order that they might be kept by themselves; and a suitable inscription was to be placed over the collection, denoting that it was the donation of Dr. Dale.

Mrs. Elizabeth Blackwell, the celebrated authoress of the Herbal, which was published about this time, resided at Chelsea opposite the Physic Garden, for the convenience of having daily access to such rare and curious plants as could nowhere else be found. As her engravings were stated to be executed after drawings taken from the life, this privilege of access to the Garden must have been of great importance to her.

It will excite no small surprise to find that the green-house, which had not been finished more than seven years, was at this time in such a state as to excite serious alarm for its safety. Such however appears to have been the case, for in the month of August, the opinion of two architects was taken upon the subject and delivered to the Court of Assistants; which opinion stated that the building required reparations to the amount of 300*l.* to prevent it from falling down. 1741

Probably the danger in 1741 was not so imminent as to require immediate attention, for nothing important was done in the matter until the present year, when the repairs executed in the Garden amounted to 215*l.* 1743

Mr. Joseph Miller, who served the office of Master in 1738, was chosen Demonstrator of plants, in the room of Mr. Isaac Rand deceased, with a salary of 34*l.* exclusive of the interest of Mr. Meres' legacy, which amounted to 8*l.* Mr. Joseph Miller does not appear to have been any relation to Mr. Philip Miller, the distinguished

Gardener of the Society, but he was a botanist of considerable reputation.

He published a very good work on Botany, bearing the following title: — "Botanicum Officinale, or a Compendious Herbal, giving an Account of all such Plants as are now used in the Practice of Physick, with their Descriptions and Virtues. By Joseph Miller, 1722." This book is dedicated to Sir Hans Sloane, and in the Dedication allusion is made to the "noble Present of the Physick Garden at Chelsea to the Apothecaries' Company on condition of its being kept up to the improvement of Botany." A great number of plants are described, the Latin name being in each case given first, and the English one afterwards, the quantities of the Latin vowels, where they are doubtful, being very accurately given. As was usual with the Medico-botanical books written at this and previous periods, a great number of plants are included which have no claim to a place in the Pharmacopœia. The arrangement is alphabetical, and no other classification is attempted, the names being adopted from Bauhin, Gerard, and Parkinson, and the English plants being described after Ray, in his Synopsis.

Mr. Joseph Miller completed another work, in two volumes, now in the Library of the Apothecaries' Hall, called "Icones Plantarum," which consists of a series of coloured figures of plants, generally very well executed, although some of them are but rough sketches. Some are so accurately delineated that they appear to be and probably are fac-similes of the plants themselves, made by pressing the leaves on paper and subsequently colouring the impressions thus obtained. This book has no date, but the influence of Linnæus is perceptible, for there is some attempt at arrangement, the Composite plants being ranged together, as are likewise the Umbelliferæ, the Filices, the Liliaceæ, the Leguminosæ, the Ranunculaceæ, the Rosaceæ, the Labiatæ, and many others. The work seems to have been all done by hand, and the names of the plants are written with a pen and ink. The book is in an excellent state of preservation, and the colours are still vivid and life-like.

1744

An order was made that no person should be permitted to gather specimens from the Garden without leave from the Director or the Gardener, and that no person whatever, who was not a Member of the Society, should be permitted to walk in the Garden without the attendance of the Gardener. The improper conduct of some individual must be presumed to have been the cause of this prohibitory regulation, for it is stated that a certain gentleman, having exceeded all bounds of reason and good manners in taking specimens from the garden, be in future excluded.

Mr. Joseph Miller produced a catalogue of Botanical books presented by the executrix of Mr. Isaac Rand.

1745 Mr. Robert Nicholls presented twelve volumes of dried plants, which were deposited in the green-house.

1746 Mr. Joseph Miller offered to reduce his salary as Demonstrator from 34*l.* to 27*l.* which with the addition of Mr. Meres' legacy amounted to 35*l.* This offer the Court accepted, though no reason is assigned either for the proposal itself or for the acquiescence of the Court.

1747 The state of the green-house now became an object of immediate attention, and the discovery must have been made that its condition was much worse than was originally supposed, for towards the close of this year the Master acquainted the Court of Assistants, that he had waited upon Sir Hans Sloane with a statement of the very extensive repairs which appeared at this time to be necessary for the Garden, and of the estimated expense, which amounted to 933*l.* 15*s.* At this interview Sir Hans offered 100 guineas towards this sum, and he promised that he would endeavour to procure further assistance from some of his friends. These generous and liberal offers the Court thankfully accepted, and the repairs immediately necessary, amounting to 175*l.*, were ordered to be commenced as soon as the season would permit.

1748 On the 21st of March of this year Sir Hans Sloane presented 150*l.* towards the repairs of the Garden. It may be presumed that this very handsome donation included the 100 guineas above mentioned. Dr. John Wilmer was appointed Demonstrator of Botany, in the room of Mr. Joseph Miller deceased, and his salary was fixed at 30*l.* per annum, including the interest of Mr. Meres' legacy.

The widow of Mr. Joseph Miller presented twenty-two volumes of dried plants, for which the thanks of the Court of Assistants were presented to her.

It was again proposed to open a subscription among the Members, on account of the great expense to be incurred in repairing or rebuilding the green-house, but, as nothing more is said of this proposition, it was probably abandoned.

1750 During the present year, the Committee reported that, having carefully examined the Garden, they found it in very good order, and that they were well satisfied with the appearance of a very large number of rare plants, many of which were non-descripts, namely, plants not yet described; and that these acquisitions to the Garden were due to Mr. Miller's great diligence in maintaining a foreign correspondence, and procuring seeds and plants from various parts of the world.

1751 The statue of Sir Hans Sloane having been lately removed from its original station in front of the green-house, and placed upon a pedestal in the Garden (where it now stands), the following inscription in Latin, prepared by Mr. Warden Chase, was approved and executed :—

Hansio Sloane, Baronetto, Archiatro,
Insignissimo Botanices Fautori,
Hoc honoris causâ Monimentum,
Inque perpetuam ejus Memoriam,
Sacrum voluit
Societas Pharmacopœiorum Londinensis.
1733.

On two other sides of the pedestal, are the following inscriptions :—

They being sensible how necessary
that branch of science is to the
faithful discharging the duty
of their Profession, with grateful
hearts and general consent,
ordered this statue to be erected,
in the year of our Lord 1733,
that their Successors and Posterity
may never forget their common
Benefactor.

Placed here in the year 1737.

Sir Benjamin Rawling, Knt. *Master.*

Mr. Joseph Miller,
Mr. Joseph Richards. } *Wardens.*

The Garden Committee which, in the year 1722 was directed to consist of the Master, Wardens, and nine other Members of the Court, was continued with the same constitution as to the number of members, but it was now ordered that three of them should be changed every year. 1752

The annual expense of the Garden, which for several years past had fluctuated between 200*l.* and 220*l.* was this year increased by an expenditure of 108*l.* in repairs. This increase of expense appears to have given rise to the following Minute of the Court of Assistants, namely: "A report of the expense of the Garden, for the last seven years, having been taken into consideration; the Master and Wardens were directed to wait on the Earl of Macclesfield, President of the Royal Society, and to represent to him the great expense the Society sustained in maintaining the Botanical Garden at Chelsea; for which Sir Hans Sloane made no provision in his will; and to consult with him as to the future measures the Society should take." This interview most probably took place, but of the result of it there is no account, nor of anything having been done in consequence of this representation. Sir Hans Sloane had died at the beginning of this year. 1753

1759

A legacy of 100*l.* was received from Mrs. Rand, in trust for the following special purposes; namely, that two-thirds of the annual interest be paid to the Demonstrator for the time being, for placing twenty newly-dried specimens of plants yearly, in her late husband's collection, in the room of such as might be decayed, and the other third of such interest to be paid to the Master and Wardens for seeing this duty performed. This legacy was ordered to be invested in 3 per cent. Annuities.

1761 It must be presumed that there were at this time no suitable apartments in the green-house for the residence of the Gardener, as there had been for many years before; for a memorial was now presented from Mr. Miller, requesting the Society to build him a dwelling in the Garden. In consequence of this memorial, Dr. Wilmer and Mr. Miller were desired to procure a proper habitation for the latter and his family, until the Court should be able to provide some permanent and more suitable accommodation. What steps of a permanent nature were taken upon this business cannot at present be ascertained.

1764 On the 25th of October of this year Dr. John Wilmer resigned his office of Præfectus Horti and Botanical Demonstrator; and on the 14th of March following, Mr. William Hudson was 1765 elected to succeed him in those offices.

Mr. Miller presented a memorial to the Court 1767 of Assistants, setting forth his services from the time of his appointment in 1722 to the present period, and also the various expenses he had incurred for the improvement of the Garden, for which he had never been indemnified; these he represented as amounting to 62*l*. and he stated moreover that his present expenses were such as to consume the whole of his salary.

In consequence of this memorial, the Court of Assistants ordered that 50*l*. be given to him as a consideration for his expenditure in correspondence, &c.; and that although his salary be not increased, such annual gratuity should be given to him as the Garden Committee shall recommend, and the Court approve.

The following document, extracted from the Minutes of the Garden Committee, and connected, doubtless, with the preceding memorial, is too curious to be omitted. It is certainly of a very singular description, and is by no means easy, at this distance of time, to be fully understood, particularly that part of it which relates to the payment of the men's wages out of the Gardener's salary. It is entitled

"*An exact account of all money received, and the money paid to workmen from Christmas* 1765 *to Christmas* 1766, *by* Mr. Miller, *Gardener, for keeping the Botanic Garden at Chelsea.*

	£	s.	d.
To money paid by the Worshipful Company of Apothecaries	50	0	0
Taken at the gate for showing the Garden .	31	4	0
	£81	4	0
Paid to 3 men, 52 weeks at 9*s.* per week .	70	0	0
—— to a fourth man at 10*s.* per week .	4	0	0
Charges of several parcels of seeds sent by the Post, and also of plants sent per sea .	15	5	0
	£89	5	0

"By which it appears that I am a considerable loser, instead of having anything for myself. And from the inclemency of the season, I have not received 4*s.* per week on an average since Christmas last."

1768 A gratuity of 50*l.* was ordered to Mr. Miller, as an acknowledgment of his various services to the Society, which were enumerated as the ground of such grant.

1769 Mr. Stanesby Alchorne, having been requested to prepare a new Catalogue of the books remaining in the Library at Chelsea Garden, presented the volume in the following year to the Committee for managing the Garden, and received their thanks.

From this Catalogue it appears that the

Library at that time consisted almost entirely of books confined to Botanical subjects, 266 in number, and of pamphlets and unbound books about 50; besides which there were, in the Library at Apothecaries' Hall, 238 volumes of books, the greater part of which were on subjects relating to Botany and other departments of Natural History, many of them having been at different times removed thither from Chelsea. The Society likewise possessed twenty volumes of dried plants, being the Hortus Siccus of Mr. Joseph Miller; and a bundle of dried plants given by Dr. John Wilmer, who succeeded the former as Præfectus Horti. Mr. Isaac Rand's collection is not here mentioned.

A Catalogue of the plants contained in the Garden, and an inventory of every particular of the Society's property there and in the greenhouse, was ordered to be made. This order was intrusted to a Committee, who were allowed to obtain such assistance as might be requisite, and considerable progress was made in their work, but in consequence of some opposition, said to be offered by Mr. Miller, great delay was occasioned; and afterwards a new Committee was appointed for the same purpose, in order to obviate the difficulties raised by Mr. Miller. Whatever these difficulties

1770

might have been, the new Committee were not more successful in their proceedings than the old one. Fresh complaints were made of Mr. Miller's conduct, and the differences terminated in a request by that gentleman to be allowed to resign the office of Gardener, which resignation was accepted by the Court of Assistants. The Garden Committee were thereupon authorised to treat with Mr. William Forsyth, who had offered himself as a candidate for the vacant office, and, after being approved, he was appointed Gardener. His salary was to be 60*l.* per annum, with lodging-rooms in the green-house, and coals. He was likewise to be allowed 50*l.* per annum for two or more under-gardeners to be provided by himself, but was prohibited from selling roots or plants. To these terms Mr. Forsyth agreed, and entered upon his duties accordingly.

Proper rules and directions for the conduct of the Gardener having been drawn up, they were entered at length in the Minute Book of the Committee (28th December, 1770), and a copy of them was ordered to be hung up in the Library.

It is greatly to be lamented that the last years of the services of such an eminent Botanist as Philip Miller, who had held his appointment almost half a century, should be overshadowed by the differences which arose between the Society and himself in conducting the affairs of the Garden. On his part these differences may have arisen solely from the infirmities and the irritability consequent upon his great age; and, on the part of the Committee, they may have been caused by too great an anxiety for the success of the Garden, a feeling influencing those gentlemen to show less forbearance with respect to his infirmities than they would otherwise have done. Under these circumstances the resignation and its acceptance were highly commendable, and the grant to Mr. Miller of a retiring pension of 50*l.* appears to have produced on both sides a cordial reconciliation. But he survived only a very short time to enjoy it.

1771

The following Biographical sketch of this distinguished Botanist cannot but be interesting to every cultivator of Horticultural and Botanical science.

Mr. Philip Miller, F.R.S.

Philip Miller was born in or near London in 1691. He raised himself, by his own merit, from a state of obscurity to a degree of eminence rarely attained by a person in his station of life, and arrived at the distinction of being considered by the most eminent authorities as the first Botanical Gardener of his time. In 1722, on the nomination of Sir Hans Sloane, he was appointed Gardener to the Botanic Garden at Chelsea, which office he filled with great reputation for the long space of forty-eight years. He was not only intimately acquainted with the indigenous vegetable productions of Britain, but had an extensive knowledge of exotic plants. He added to his practical knowledge that of the structure and classification of plants, and was well acquainted with the writings of Ray and Tournefort. Habituated to the systems of these two great Botanists, it is not surprising that he was somewhat reluctantly brought to adopt the novel views of classification introduced during his lifetime by the illustrious Linnæus. But so learned and intelligent a man as Philip Miller could not fail to be struck with the profound sagacity and the far-reaching views of the Swedish naturalist, and we accordingly find that he not only renounced his allegiance to former authorities, but became an enthusiastic advocate of the new doctrines. In his "Short Introduction to the Knowledge of the Science of Botany, explaining the terms of Art made use of in the Linnæan System," published in 1760, Philip Miller writes as follows :—

"Doctor Linnæus, the celebrated Professor of Botany at Upsal in Sweden, has prepared simple and proper terms, not only to express all the different parts of plants, but also the principal qualities, forms, figures, situations, directions, and manners of existing of each of these parts . . . . These principles have been almost universally adopted by those who have wrote after him. . . . . The sexual method of classing plants, established by Doctor Linnæus, is much preferable to all the systems of Botany which have yet appeared. 1. Because of all those which have been proposed, there is not one of them which approaches so near to a natural method as this, most of the classes being very natural. 2. It is founded upon the parts of the plants which are the most constant, and least subject to variation, the stamina and pointals, which are the true organs of generation."

The first edition of his great work, "The Gardener's Dictionary," was published in 1724, and it afterwards appeared in 1731 in a folio volume; the seventh edition was published in 1759, and in this the author adopted for the first time the system of Linnæus. Besides these editions, several others in an abridged form were published in octavo, and one appeared in quarto.

The seventh edition of the Dictionary is

generally esteemed the most complete, and is entitled "The Gardener's Dictionary; Containing the best and newest Methods of Cultivating and Improving the Kitchen, Fruit, Flower Garden, and Nursery: As also for performing the Practical Parts of Agriculture; Including the Management of Vineyards, with the Methods of Making and Preserving the Wine, according to the present practice of the most skilful Vignerons in the several Wine Countries of Europe. Together with Directions for Propagating and Improving, from real practice and experience, all sorts of Timber Trees. The seventh edition, revised and altered according to the latest system of Botany, and embellished with several copper-plates, which were not in the former editions. By Philip Miller, F.R.S., Gardener to the Worshipful Company of Apothecaries, at their Botanic Garden at Chelsea, and Member of the Botanic Academy at Florence.

'Digna manet divini gloria ruris.'—*Virg. Georg.*

London, MDCCLIX." The eighth edition was published in 1768, and was the last which appeared in the author's lifetime.

The merit of this work is too well understood and recognized to render any detailed description necessary. It is sufficient to state that Miller's

Dictionary has been translated into the Dutch, German, and French languages, and that a new and greatly improved edition of it, with considerable additions, was published by Dr. Thomas Martyn, Regius Professor of Botany in the University of Cambridge, in the year 1807, in four folio volumes. The Botanical and Horticultural information contained in these four large tomes is enormous, and even in the present day the work must be considered a prodigy of science, learning, industry, and research, as well as of practical skill.

Philip Miller wrote many other, but smaller works, and contributed many papers to the learned Societies of his time. Allusion has already been made to his Catalogue of the Plants of the Chelsea Garden, and to his "Short Introduction to the Science of Botany," the latter being an excellent little work on the principles of the Linnæan nomenclature and classification. He wrote also the "Gardener's Kalendar," which directs what works are necessary to be done every month in the Kitchen, Fruit, and Pleasure Garden, as also in the Conservatory and Nursery. The tenth edition appeared in 1754, and it is inscribed to the Master, Wardens, and Court of Assistants of the Apothecaries' Company.

Mr. Miller passed some time in Holland in order to acquire a knowledge of the practice of the famous Florists of that country. The improvements he introduced into Horticulture were many and great, and he maintained a correspondence with the most eminent foreign Botanists, and particularly with Linnæus, who said of his Dictionary, "Non erit Lexicon Hortulanorum, sed etiam Botanicorum." By other foreigners he was styled "Hortulanorum Princeps." He was a Member of the Botanic Academy at Florence, and a Fellow of the Royal Society of London, being occasionally elected one of their Council.

He had at various times several pupils under him, who afterwards attained eminence, and among them are to be enumerated Aiton and Forsyth. The latter succeeded him in the care of Chelsea Garden. Mr. Miller's great age and infirmities, together with other circumstances to which allusion has already been made, induced him to resign the office of Gardener a short time before his death, which took place at Chelsea, on the 18th December, 1771. He left a very large Herbarium of exotic plants, collected principally from the Chelsea Garden, and this valuable collection was purchased by Sir Joseph Banks. Mr. Miller was never a rich man, and he left very little money, for he was of too generous and careless a disposition to accumulate wealth. He had, however, sufficient opportunities of doing so, for, besides the profits of his published works, he was consulted by various noblemen and gentlemen on the subjects of planting, laying out grounds, &c.; particularly by the three noble Dukes of Bedford, Northumberland, and Richmond. The small retiring pension awarded to him by the Society of Apothecaries, to which he was amply entitled by his long and distinguished services, must have been enjoyed by him for a very brief period indeed, for it was granted in 1770, and he died in 1771.

A small genus of plants of the Natural Order of Compositæ, and discovered at Panama and Vera Cruz, is called *Milleria*, this name having been given to it by Linnæus, in honour of Mr. Miller.

Dr. Pulteney says that Mr. Miller was the only person he ever knew who remembered having seen Mr. Ray, and that he should not easily forget the pleasure which enlightened his countenance when he related the fact; it so strongly expressed the "Virgilium tantùm vidi," when, in speaking of that revered man, Miller referred to this incident of his youth.

Philip Miller married the sister of Ehret, the

famous Botanical Painter, and had two sons, Philip and Charles. The former, after having worked some years under his father, went to the East Indies, where he died. The latter, Charles, was educated at Chelsea, and was appointed by Dr. Walker, the Founder of the Botanic Garden at Cambridge, as his first Curator. After remaining some years at Cambridge, he also went to the East Indies, whence he returned after many years with an ample fortune, and became a resident in London. He was a good Botanist, and there are some papers, written by him, in the Philosophical Transactions. He died October 6th, 1817, at the age of seventy-eight, and was buried at Chelsea in the same grave with his father.

Nearly half a century subsequent to the decease of Philip Miller, the Members of the Linnæan and Horticultural Societies, sensible of his great talents, determined, very highly to their honour, to leave some record of his merit, and to make it extensively known by erecting a monument to his memory, the expense of which memorial was to be defrayed by a private subscription among themselves.

This project was accordingly carried into execution, and a monument was erected, not immediately over the place of his interment, the situation of which would not conveniently admit it, but in a more conspicuous part of the church-yard. It is a Cenotaph in the pillar form, the pedestal being circular, and the upper part surmounted by an urn enriched with foliage; and the whole is surrounded with an iron railing.

On the pedestal is the following inscription:

PHILIP MILLER,
sometime Curator of the Botanick Garden,
Chelsea;
and Author of the Gardener's Dictionary,
died December 18th, 1771, aged 80,
and was buried on the North side of
this Church-yard,
in a spot now covered by
a Stone inscribed with his Name.
The Fellows of
the Linnæan and Horticultural Societies
of London,
in grateful Recollection of
the eminent Services rendered to
the Sciences of Botany and Horticulture
by his Industry and Writings,
have caused this Monument to be
erected to his Memory.
A.D. 1815.

In the summer of this year, it was agreed, 1771 with the consent of the Water-Bailiff on the part of the City of London as Conservators of the river, to embank the Garden towards the Thames, and this embankment was carried into effect at an expense exceeding 400*l*. The first brick of this work was laid on the 7th of June, in the presence of the Committee, by John Lisle, Esq., Master of the Society, who placed under it two pieces of his Majesty's coin of the present year. This embankment was designed only in order to recover ground which had originally belonged to the Garden, but had in process of time been washed away by the river.

In the spring of this year Mr. William Hudson resigned the office of Demonstrator of Plants. Mr. Stanesby Alchorne, a Member of the Society, and Assay Master of his Majesty's Mint, was requested to supply his place, and kindly undertook the duty, *ad interim*, until another person should be elected. It was at the same time recommended to him to introduce a more scientific method at the Botanical Lectures than had been previously practised.

Mr. William Hudson, F.R.S.

William Hudson was a man whose rank as a Botanist was too high to allow him to be passed over in silence in these memoirs.

He was born at Kendal, in the county of Westmoreland, about the year 1732. He served his apprenticeship with an Apothecary in Panton Street, whom he afterwards succeeded in business. The amiable and learned Benjamin Stillingfleet was his early friend in the study of Natural History, and directed his attention to the writings of Linnæus with such effect that Hudson may be justly considered as having been one of the earliest Linnæan Botanists in England, and he was probably the first author in this country who embraced that system, Philip Miller having adopted it subsequently.

Hudson's "Flora Anglica" was first published in 1762 in one volume, and a second edition in two volumes appeared in 1778. The elegant Preface to the first edition is said to have been from the pen of Stillingfleet.

The book is entitled "Gulielmi Hudsoni, Regiæ Societatis Socii et Pharmacopœii Londinensis, FLORA ANGLICA; exhibens Plantas per Regnum Britanniæ sponte crescentes, distributas secundum Systema Sexuale, cum Differentiis Specierum, Synonymis Auctorum, Nominibus Incolarum, Solo Locorum, Tempore Florendi, Officinalibus Pharmacopœiorum." (The ENGLISH FLORA of William Hudson, Fellow of the Royal Society, and Apothecary of London; describing the Plants growing wild in the Kingdom of

Britain, and arranged according to the Sexual System; together with the Differences of the Species, the Synonyms of Authors, the Names given by the Inhabitants, the Soil of the Localities, the Time of Flowering, and the Officinal Plants of Apothecaries.) This title sufficiently indicates the contents of the book, which is very creditable to Mr. Hudson's learning and to his practical skill as a Botanist. It is written altogether in Latin, with the exception of giving the English names of the plants. In the arrangement of the materials Hudson acknowledges his obligations to the Synopsis of Ray, and he states in his Preface, that if the Science of Botany had received no additions since the times of that great Naturalist, nothing more would be wanted for distinguishing British plants except a new edition of his Synopsis, which, as Hudson remarks, was then out of print. But he goes on to say, "exortum est his diebus novum sidus, quod orbi Botanico lucem affudit ne in somnis quidem antea visam: magnus scilicet ille Linnæus, qui partes plantarum minutissimas, vel prius neglectas, vel non omnino observatas in conspectum producendo, plene illustrando, et rite nominando fundamentum jecit novæ methodo simplici, compendiosæ, certæ: quam et constantiâ plane admirandâ toti regno vegetabili adhuc cognito feliciter applicuit." (A new star has arisen in our days which has shed a light, not before seen even in dreams, on the Botanical world; namely that great Linnæus, who, by bringing to view, by fully illustrating, and by correctly naming, the most minute parts of plants which had either been previously neglected or had never been observed, has laid the foundation for a new and simple, concise, and certain method, which he has happily applied, with an admirable consistency, to the whole vegetable kingdom hitherto known.)

The method of Linnæus is therefore followed in all respects by Hudson in his "Flora Anglica," the plants being all arranged in the Linnæan classes and orders. The cultivation of indigenous Botany had long been a favourite pursuit in England before Hudson's time, and the "Flora Anglica" was no doubt a welcome addition to the existing knowledge of the subject. The descriptions are very clear, the collection of synonyms is very copious, the places of growth of the different plants are carefully indicated, and references to the Pharmacopœia are given in the case of the medicinal plants. More than half of the second volume (in the second edition) is occupied by the Cryptogamic Plants, comprising the Ferns (Filices), the Mosses (Musci), the Sea Weeds (Algæ), and the Mushrooms (Fungi); and this part of the work must have been a very difficult task, the accurate descriptions, definitions, and distinctions of these forms of vegetation involving some very laborious problems, even in the present day. It cannot therefore be a matter of surprise that Hudson admits in his Preface that the Cryptogamic class was in his time an obscure one, and had caused him very great trouble. (Classis Cryptogamia, quæ ad hæc usque tempora, quantum scio, obscura manet, et mihi plurimum molestiæ creavit.)

This publication gave Mr. Hudson considerable reputation as a Botanist, both at home and abroad. He had formed some very valuable connexions among the cultivators of Natural History, and was the correspondent of Linnæus, Haller, and other eminent foreign Physicians and Naturalists. He was elected a Fellow of the Royal Society on the 5th of November, 1761. He took an active part in the affairs of the Society of Apothecaries, particularly as one of the Committee for the management of Chelsea Garden, and in 1765 he was appointed Præfectus Horti and Botanical Demonstrator to the Society, but he resigned those offices in 1771. In the winter of 1783 his house and the greater part of his valuable treasures were destroyed by fire, which was generally believed to have been caused designedly by a servant who knew of a considerable sum of money which Hudson had received a day or two before. Being at the time uninsured, his loss was very great, and it was not only severely felt by a man of his limited resources, but the calamity at the same time defeated a project he had entertained of publishing a Fauna Britannica, for which he had long been collecting materials.

After the fire he removed to Jermyn Street, and, having retired from practice, and having never married, he resided with Mr. and Mrs. Hole, the latter of whom was the daughter of the gentleman to whose business he succeeded. In 1791 he became a Member of the Linnæan Society. He died in Jermyn Street on the 23rd of May, 1793, and was buried in St. James's Church, Piccadilly. He bequeathed to the Apothecaries' Society his dried specimens of plants. A rather rare genus of plants belonging to the Natural Order of Cistaceæ, was named *Hudsonia* by Linnæus in honour of Mr. Hudson.

At the recommendation of the Committee, the two northern cedar-trees, being in a decayed state, were cut down, together with several lime and elm-trees, and some others in the officinal quarter, which were considered injurious to the growth of the other plants for which the Garden was more particularly designed. Such a number, however, were left as were esteemed consistent with the beauty, well-being, and ornament of the whole Garden.

During the present and following year a great interchange of exotic plants took place between the Society and the following Noblemen, Gentlemen, and others; namely, from H.R.H. the Princess Dowager's Garden at Kew; His Grace the Duke of Northumberland's at Sion; Mr. Ord's at Walham Green; Dr. William Pitcairn's at Islington; Mr. Gordon's Nursery Garden at Mile End; Mr. Lee's Nursery Garden at Hammersmith; Mr. Watson's Nursery Garden at Islington; Dr. John Fothergill's Garden at Upton; the Rt. Hon. the Earl of Coventry's; Richard Warner's, Esq., at Woodford; Mr. Bewicke's at Clapham. A bag of seeds was presented by Joseph Banks, Esq., and Dr. Solander.

It may be recorded as a circumstance of some

interest, that the produce and value of the two cedar-trees which had been cut down were, of the trunks 133¾ feet at 2*s*. 8*d*. per foot, and of the boughs, 84¾ feet at 1*s*. 4*d*. per foot, amounting altogether to the value of 23*l*. 9*s*. 8*d*. The two remaining cedar-trees, with their stately branches meeting together, formed for many years a conspicuous object as seen from the river, but as has before been mentioned, one of them has lately died and the other is dying (1878).

1772

Mr. Stanesby Alchorne, Honorary Demonstrator, produced the twenty dried plants which he had added to Mr. Rand's collection in the present year, agreeably to the will of the late Mrs. Rand. It must be presumed that this condition had been hitherto regularly complied with, though no previous notice occurs in the Minute Books concerning it. Mr. Alchorne also placed in the Garden nearly fifty new trees, particularly of the Genera *Cratægus*, *Mespilus*, and *Pinus*, which he presented to the Society, and the thanks of the Committee were given to him for the donation.

In addition to this donation he presented about forty tons of old stones, brought from the Tower of London, for the purpose of raising an artificial rock, in order to cultivate those plants which require such a soil, and to this rockwork was afterwards added a large quantity of flints and chalk given by Mr. John Chandler, and also a large quantity of lava from a volcano in Iceland, presented by Joseph Banks, Esq. These materials being considered fully adequate to the purpose, the rockwork was undertaken, and was finished in the course of the summer of the following year.

It was ordered by the Court of Assistants, that each Member of the Garden Committee be allowed 2*s.* for coach-hire upon each attendance, and that the Committee should not exceed 220*l.* per annum in the expenditure upon the Garden without the consent of the Court.

Mr. Alchorne, having officiated as Honorary Demonstrator of Plants about two years, since the resignation of Mr. Hudson, now resigned his office, and was presented with a piece of plate, value thirty guineas, as a small acknowledgment for his services.

1773 On the 15th of December, Mr. William Curtis was elected to the vacant office of Demonstrator of Plants and Præfectus Horti. The duties of this office were set forth at large by the Committee in the following terms:—

"1st. The office of Botanical Demonstrator to this Society, is to superintend their Garden and Gardener, as also their library, and all other matters upon their premises at Chelsea; but with submission always to the superior direction of the General Committee for the management of the Society's Garden. His duty is to encourage and cultivate the knowledge of Botany, as well theoretic as practical, among the students of this Society; for which purpose,—

"2ndly, He is to attend the Society's Garden at stated times, not less than once in every summer month (from April to September, both inclusive) to demonstrate the plants, especially in the officinal quarter, with their names and uses. The last Wednesday in each of the above months has been usually appropriated to this service, beginning at nine of the clock in the morning.

"3rd. He is expected to make some annual excursion, for two days at least, preparatory to the Society's General Herborizing, inviting two or three of the ablest botanical members to his assistance, the intention being to collect such vegetables as are not commonly found in the environs of this metropolis, to be demonstrated by him at the meeting appointed for that purpose, and he will receive 3*l.* towards defraying the expensee of very such journey, pursuant to the will of Mr. Robert Loggan, a late worthy member.

"4th. He is to accompany and conduct the Students of this Society in their search after indigenous plants, upon every day appointed for their private herborizings, which are only five in each summer; when he is desired to use his best endeavours in preserving strict decorum among his pupils, and in directing and confining their attention solely to the intended business of the day. And as the regular lectures lately introduced on these occasions at the request of the Master and Wardens, have, on account of the times and places, appeared insufficient for teaching accúrately the elements of botanic science, it is now recommended for the demonstrator to consider of some more effectual method of answering so desirable an end.

"5th. He is yearly to prepare fifty dried specimens from plants, growing in the Society's Garden at Chelsea, which are to be presented to the Royal Society, by direction of the late Sir Hans Sloane, Bart., having been first approved by the Court of Assistants of this Society. Also to dry twenty other specimens, in lieu of so many plants which shall be found decayed in the collection of the late Mr. Rand, now in the library at Chelsea. These to be placed in the said Herbarium before the first day of May in every year, and there will be fifty shillings paid him for every such service, by appointment in the will of the late Mrs. Rand.

"6th. He is to attend each private Court, at the Hall during the summer months, to give his advice (if required) relative to the private herborizings; as likewise to be present at the several meetings of the General Committee for managing the Garden, where it has been customary for him to act as Secretary. And in this capacity it is earnestly recommended to him to cultivate an extensive botanical correspondence, both at home and abroad; and to keep copies of the same, as the most probable means for propagating the knowledge of this science; for enlarging and improving the Garden under his inspection; and for promoting the interest and honour of this Society in general."

These instructions were directed to be entered in the General Order Book at Chelsea, and a copy of them was to be delivered to Mr. Curtis for his private inspection. The Beadle of the Society was likewise ordered to give proper notice to the Demonstrator of every intended meeting of the Committee. During the present year various donations of seeds and plants were received from several of the noblemen and gentlemen before mentioned; and also from Dr. Young, of St. Vincent's; Captain Blake (seeds transmitted from his son in China); Dr. Hope, of Edinburgh; Mr. Malcolm, of Kennington; Mr. Daniel Mildred.

Mr. Curtis's proposal for giving lectures on Botany, at the Hall in Blackfriars, for the purpose of explaining systematically the Principia Botanica, was approved and recommended to be carried into effect on the first and third Wednesdays in the summer months, at nine in the morning, but, for some reason which does not now appear, this plan was not carried into effect.

1774 During several days in the month of March very high tides occurred in the Thames, in consequence of which the water rose fifteen inches within the nursery in the Chelsea Garden, although every endeavour was made to prevent the effects of the accident by damming up the gates and by making trenches to carry off the water.

Mr. Forsyth, the Gardener, having complained that his salary was insufficient, the Committee proposed to recommend to the Court of Assistants, that, in order to augment it, he be allowed to dispose of supernumerary specimens of plants, under proper restrictions, for his own benefit, and this arrangement was accordingly permitted. Such a permission, however, is of so objectionable a nature, that it is difficult to conceive how the managers could have been induced to sanction it.

The subjoined statement of the expenses of the Garden, from the year 1741 to the present year, is copied from a paper drawn up by Mr. John Field, an active and intelligent member of the Society in this and other departments. 1775

From this statement it will appear that the average annual charge, exclusive of the extraordinary expenses of those years in which important repairs or erections were made, did not exceed 240*l*. This calculation will be applicable only to the first thirty years of this period, subsequently to which the annual charge was rapidly increasing.

	£	*s.*	*d.*	
1741	289	14	1	
1742	195	12	5	
1743	355	1	9	Repairs £196
1744	238	4	11	
1745	163	19	11	
1746	212	1	0	
1747	245	19	7	
				£1700 13 8

	£	s.	d.				
1748	388	6	9	Repairs £175			
1749	221	10	7				
1750	234	11	1				
1751	210	19	11				
1752	202	19	4				
1753	322	18	7				
1754	200	16	11				
					£1782	3	2
	£	s.	d.				
1755	197	17	4				
1756	213	6	7				
1757	278	14	5				
1758	219	1	5				
1759	215	4	8				
1760	252	10	5				
1761	233	16	6				
					£1610	11	4
	£	s.	d.				
1762	204	19	3				
1763	256	15	8				
1764	240	13	4				
1765	195	15	9				
1766	221	17	3				
1767	226	9	9				
1768	322	10	0				
					£1669	1	0
	£	s.	d.				
1769	282	11	11				
1770	269	5	6				
1771	917	16	8				
1772	383	2	5				
1773	291	15	0				
1774	306	4	3				
1775	339	13	4				
					£2790	9	1
Total amount in thirty-five years . .					£9552	18	3

The very large expenditure of the year 1771 arose from the expense of embankment, and other extra charges connected with this work, as well as from necessary repairs.

The Court of Assistants directed that every Member of that Court, who had served the office of Master, be a standing Member of the Garden Committee, and that each Member of that Committee be allowed 5*s.* for every attendance. 1776

Mr. Forsyth, the Gardener, having complained of the insecurity and inconvenience of his residence, the Committee were directed to examine into the subject and to consider what alterations might be made to improve the building; upon which an estimate was presented of such improvements as appeared necessary, amounting to 87*l.*, exclusive of iron work. This estimate was adopted, and the alterations were carried into effect; and to the expense incurred there was afterwards added 9*l.* for rebuilding two decayed buttresses in the green-house.

On the 27th of August, Mr. William Curtis resigned his office of Botanic Demonstrator, and on the 18th of March following Mr. Thomas Wheeler, F.L.S. was appointed to succeed him, with an additional allowance of 2*l.* for the July Herborizing excursion. 1777 1778

The merits of the former Demonstrator, as

a Botanist, justly demand some biographical tribute to his memory, and the following record will be read with satisfaction by all cultivators of botanical science.

William Curtis.

William Curtis was born at Alton, in Hampshire, in the year 1746. His father was a tanner in that town, and in the vicinity of it young Curtis received his education, which was of a very limited nature. He was placed as an apprentice to his grandfather, who practised as an Apothecary in the same place. He was early led into the study of Botany in the following rather curious manner. He happened to be residing next to the Crown Inn at Alton, the ostler of which, John Lagg, a steady and sober man, had acquired, by means of the well-known works of Parkinson and Gerarde, a considerable knowledge of plants; and his taste in this direction so forcibly impressed the mind of young Curtis, as to lay the foundation for a development in the latter of that love of Natural History which afterwards made him so deservedly conspicuous. At the expiration of his apprenticeship he came to the metropolis, and entered into the service of Mr. Thomas Talwyn, a medical practitioner in Gracechurch Street. Mr. Talwyn was a Member of the Society of Friends, to which religious persuasion the parents of Curtis also belonged. After some time the latter succeeded his employer in business, but his botanical pursuits interfered too much with professional duties to allow him to obtain an extensive medical practice, for which there is reason to believe he was otherwise well qualified.

The office of Demonstrator of Botany to the Society of Apothecaries being vacant in the year 1773, Mr. Curtis, who had obtained the favourable opinion of many persons eminent for their knowledge of Natural History, and among the rest Mr. Alchorne, was strongly recommended to the Society by that gentleman as a proper candidate to fill up the vacancy, and he was accordingly elected to the office, having previously been admitted a Member of the Society. He continued in this post about five years, when his other avocations induced him to resign it. He had united to the study of Botany that of Entomology, and in 1771 he published "Instructions for the collecting and preserving Insects," and, in the following year, a translation of Linnæus's Fundamenta Entomologiæ.

Soon after his appointment as Botanical Demonstrator to the Society of Apothecaries, he commenced lecturing publicly on Botany, both as to the principles and practice of that science. For the use of his pupils at these lectures, he occupied a piece of ground as a garden, first in the Grange Road, afterwards in Lambeth Marsh, and subsequently at Brompton, which last garden he continued to cultivate until his death, the soil and situation being there much more suitable to his purpose than at Lambeth. The principal object he had in view in cultivating these gardens was the growth of the indigenous plants of Britain.

In 1777 he commenced his great work the "Flora Londinensis," which was extended to six Fasciculi, of seventy-two plants each. The author originally intended that the plan should embrace all the plants growing within ten miles of London; and the work was conducted, so far as it went, with the greatest care and accuracy. But the sale of it, which never exceeded 300 copies, was not equal to the expense, and the undertaking was therefore necessarily abandoned, a result which in a national and scientific point of view is much to be regretted.

The full title of this work was "Flora Londinensis, or Plates and Descriptions of such Plants as grow wild in the Environs of London, with their Places of growth, their times of flowering, their several names according to Linnæus and other authors; with a particular description of each Plant in Latin and English, to which are added their several uses in Medicine, Agriculture, Rural Economy, and other Arts. By William Curtis, Demonstrator of Botany to the Company of Apothecaries. In 6 vols., London, 1777."

These handsome volumes are in the highest degree creditable to the botanical acquirements of Mr. Curtis and to the skill of the artists whom he employed. The descriptions of the plants are placed opposite to the coloured plates, in parallel columns of Latin and English, and it is impossible to speak too highly, even now, after an interval of more than a hundred years, of the beauty and accuracy of the illustrations. Being executed on large folio sheets they represent the plants, in most instances, of their natural dimensions; the colours are as fresh and bright as if the objects were painted in the present day, and the artists have not only imitated exactly the structural peculiarities of the specimens, but they have gracefully represented what may be called the natural habit of each, and the drooping, or waving, or curving, or other characters of leaves, or stems, or flowers, are all depicted with the most finished taste and skill. Although it is perhaps not to be wondered at that so expensive a work, as this must have been, did not receive sufficient pecuniary encouragement, the six volumes which exist constitute a splendid specimen of the botanical publications of England in the last century.

But although this undertaking failed, Mr. Curtis very happily projected another, which although it was in many respects inferior to the former, yet it was more captivating to the public at large. The "Botanical Magazine" immediately became popular. It commenced its existence about 1787, soon attained a monthly sale of 3000 numbers, and was steadily continued until the death of the founder. Mr. Curtis had neither fortune nor patrimony, but he here found a valuable estate, although the continued produce of it depended entirely upon the regularity of publication.

After Mr. Curtis's death, the "Botanical Magazine," was continued under the management of Dr. Sims; and after him by Sir William Hooker; and after the death of the latter by his son, the present Sir Joseph Dalton Hooker, who still conducts it. 103 volumes of this valuable and popular work have already appeared, each containing from sixty to seventy plates, and the work fully sustains its former reputation.

Mr. Curtis's friendships were very numerous,

and among his most esteemed patrons and friends may be enumerated the names of Sir Joseph Banks, Dr. John Sims, (who continued the "Magazine,") Dryander, Withering, Hunter, Dickson, Sibthorp, and Lightfoot.

Besides the published works already mentioned there are several minor publications of his, as also some papers in the Transactions of the Linnæan Society. A large tree, growing in Africa, from which the Hottentots and Caffres make the shafts of their javelins, is named after him *Curtisia*.

Mr. Curtis laboured for nearly a twelvemonth before his death under a disease of the chest, which put a period to his life on the 7th of July, 1799, in the 53rd year of his age. He was buried at Battersea, and left behind him the character of an excellent Botanist, both practical and theoretical, an honest and friendly man, a lively, entertaining companion, and a good teacher, always ready to encourage novices, and to render the science of Botany as attractive as possible.

During this summer, above 100 packets of seeds were presented to the Apothecaries' Society from Joseph Banks, Esq. (afterwards Sir Joseph Banks), and Dr. Ryan of Santa Croix.

The sum of 25*l.* was paid to Mr. Forsyth, as an indemnity for the expenses he had incurred from the bad state of his apartments.

1779 The little stove in the Garden being too much decayed to be capable of repair, a new one with several improvements was built, the estimated expense being 49*l.* 11*s.* 10*d.*

1780 The Committee for the management of the Garden were allowed to expend 5*l.* for refreshments at each of their meetings. It was likewise ordered that they do not exceed 250*l.* in the expenses of the Garden for the ensuing year.

1781 The liberal benefactor of the Society, Sir Joseph Banks, who is said to have commenced his botanical studies in the Garden under the tuition of the venerable Philip Miller, presented more than 500 different kinds of seeds, collected in his recent voyage round the globe. Mr. Alexander Anderson also presented above 100 packets of seeds from St. Lucia.

1782 The Committee reported to the Court of Assistants, that they had considered Mr. Forsyth's plan for reducing the charges of the Garden, and found, that, if adopted, it would not save more than thirty or forty pounds annually, and as they considered the maintenance of the Garden in a respectable manner to be necessary for the honour of the Society, they judged it right to propose some plan of augmenting the income of the Corporation in order to enable them better to support this expense. This object they thought would be best attained by charging the rents, paid by the two commercial stocks, with such a certain annual proportion of their profits as might be a reasonable compensation for the lands and other advantages enjoyed by them from the Corporation, and which payment, varying according to the profit of each year, would be more equitable than a fixed annual sum. This plan was approved by the Court of Assistants, and the Clerk of the Society was ordered to give the necessary information to the Managers of the Navy and Laboratory Stocks.

From subsequent proceedings it appears that this scheme, though approved by the Court on the behalf of the Navy Stock, was not agreed to either by the Committee or by the proprietors of the Laboratory Stock. These two latter consented only to add 25*l.* per annum to their allowance for the Garden, which proposal, after being considered by the Court, was declined.

1784

Mr. Thomas Wheeler, the Botanical Demonstrator, commenced a series of lectures at the Hall, on the Principles of Botany, and carried them on for about two years, but as he did not meet with due encouragement in the attendance of pupils they were then discontinued.

Mr. William Forsyth, the Gardener, gave in his resignation in the spring of this year, in consequence of having been appointed to a similar situation in His Majesty's Garden at Kensington. He received the thanks of the Committee for his great care of the Garden while he was in the Society's service.

Mr. William Forsyth.

William Forsyth, F.A.S. and L.S. was born at Old Meldrum, Aberdeenshire, in the year 1737, and came to London in 1763, when he became a pupil of Philip Miller in the Garden of the Apothecaries' Society. He succeeded his instructor in his office of Gardener, in 1771, having been some time before that period in the service of the Duke of Northumberland at Sion. He continued, until his death, July 25th, 1804, to occupy the post of Chief Superintendent of the Royal Gardens at Kensington and St. James's.

Mr. Forsyth, having for many years given attention to the cultivation of fruit and forest trees, and to the diseases and injuries to which they are liable, succeeded in preparing a composition which was calculated to remedy those evils. In 1789, the success of his experiments attracted the notice of the Commissioners of the Land Revenue, and in consequence of their recommendation he obtained a Parliamentary reward for his discovery. Though the originality of his invention was doubted in some quarters, and occasioned considerable controversy, yet the issue of the inquiries made was upon the whole favourable to him. He published in 1791 his "Observations on the Diseases, Defects, and Injuries of Fruit and Forest Trees," and in 1802 the final result of his labours appeared in a "Treatise on the Culture and Management of Fruit Trees," which was sufficiently esteemed to pass through three editions in a very short time. A genus of plants, belonging to the Natural Order of Oleaceæ, is named *Forsythia* after this botanist.

The Committee for Chelsea Garden being at this time composed of Members of the Court of Assistants only, it was judged that the addition of some members of the Livery would be useful, and three of the latter body were accordingly appointed on the Committee.

Mr. John Fairbairn was elected to the office of Gardener, and he was directed to make the apartments over the green-house his residence, and he was requested to comply with the rules and directions for his conduct which were then read to him.

1785 Some considerable repairs were carried into effect during this summer, particularly the new slating of the green-house, repairing the glass-work of the various lights, and also the Gardener's apartments, the expense of all which amounted to about 140*l.* The tan-stove also being greatly decayed, a new one was erected in its place upon the most improved plan, at an estimated charge of 180*l.*

In consequence of this great expense, and the numerous reparations now required for the various buildings in the Garden, the propriety of asking assistance from the Society at large was taken into consideration; when it was resolved that a subscription for that purpose should be opened forthwith. This determination was communicated to the Members at the usual Common Hall held in the month of October, and being approved, it was acted upon accordingly.

It is a most gratifying feature in the history of the Garden, that the Members of the Society have stepped forward with the utmost alacrity, on every occasion when pecuniary exertions were necessary for the support of this valuable establishment. Their zeal and energy were at no time more conspicuous than at the present time, for in the following summer the Court of 1786 Assistants were informed that the subscription for the Garden had been so successful as to amount to 539*l.* 19*s.*, and that, after discharging the several workmen's bills, the remainder had been sufficient to purchase 530*l.* in the 4 per cent. Bank Annuities, as a fund for future buildings.

To facilitate the free admission of Members of the Society into their Botanic Garden, and to enable the Gardener and his servants to distinguish such Members from strangers, an engraved card suitably ornamented, and endorsed by the Master and Wardens for the time being, was directed to be given to every present and future Member, to which document the Gardener and his servants were to pay due regard: but it was understood that all persons visiting the Garden were to be attended by the Gardener, or one of his servants, and that the plants should not be taken away or damaged.

A quantity of loam was procured for the use 1787 of the Garden from Sion, with the consent of the Duke of Northumberland, and several loads of black mould from Wimbledon, with the approbation of Earl Spencer.

1789 Two new stoves were erected in the Garden at an estimated expense of 176*l.* 11*s.*

1790 Several donations of seeds were received this year from the following gentlemen; namely, Sir Joseph Banks, Bart.; James Edward Smith, M.D.; Mr. James Dixon, of the British Museum; William Pitcairn, M.D.; Professor Jacquin, of Vienna; Mr. Hare, St. James's Street; Mr. Chapman, at Mr. Secretary Stephens, Fulham; Mr. William Hudson; Commodore Gardner, of the Admiralty.

It was now proposed to erect two new tan-pits in front of the little tan-stove, and new glass cases in the room of the old tan-pits now fallen into decay. This proposal was approved by the Committee, but as they desired to proceed with caution, they directed that only one of them should be built at present, in order that they might have the opportunity of ascertaining the efficiency of one, before they incurred the expense of both, which would amount to about

1791 70*l.* This first erection gave so much satisfaction that in the following year the plan was completed by the addition of the second tan-pit.

1792 A new dry stove was erected at the west end of the green-house, in the room of the old one, which was in a ruinous state. The estimated expense was 170*l.*, 9*s.* exclusive of the charge of 12*l.* for a leaden cistern to be placed in it.

The following valuable donations are recorded 1793 in the present year; namely, Commodore Gardner, seeds from Port Jackson and Norfolk Island; Dr. James Edward Smith, bulbs and seeds from Sierra Leone; Robert Sherson, Esq; seeds from Port Jackson; Dr. Buxton, bulbs and seeds from the Cape of Good Hope; Mr. Willis, seeds from Africa; Mr. Salisbury, of Leeds, seeds from Madrid; Mr. Dixon, of Covent Garden, seeds from Germany.

The salary of the Botanic Demonstrator, which 1796 had been for many years 37*l.* 10*s.*, was increased to 45*l.* 10*s.* per annum.

Mr. John Tutton, the proprietor of the wall on the south-west boundary of the Garden, having agreed to cede his right to that wall for the term of seventy-two years in consideration of the Society paying the expense of rebuilding it, amounting to 6*l.* 12*s.*, his proposal was acceded to, under the advice of the Society's Solicitor, and on the 22nd June, 1797, the Corporation seal was affixed to the Lease.

Some encroachment appears in this year to 1801 have been made on the Barge-house, which, however, must have been of a very trivial nature, for it was ordered at a Court of Assistants

(30th January) that 2*s.* 6*d.* should be paid for this encroachment, and the matter was settled.

1804 Two new water-troughs lined with lead, for the cultivation of aquatic plants, were provided at a cost of 24*l.*, and placed in a convenient part of the Garden near the centre.

1806 The dried plants bequeathed to the Society by the late Mr. William Hudson, were presented to Mr. Thomas Wheeler, the Demonstrator.

The two Barge-houses were let on Lease to the Goldsmiths' Company for twenty-one years at the net annual rent of 10*l.* for each Barge-house, the tenants to make all repairs and alterations at their own expense.

1807 It was directed by the Committee, that no person be allowed to borrow any book from the Library without their leave, and that no book be retained longer than three months, and the Library was to be annually inspected at the Committee meeting in the month of August.

1808 Lord Valentia was permitted to be supplied with plants of the true Ginger (*Amomum Zinziber*), Turmeric (*Curcuma longa*), and Arrow Root (*Maranta arundinacea*).

1809 In this year the cedar-trees were much damaged by the heavy falls of snow which took place during the previous winter. Several rare plants were presented by James Vere, Esq., from the collection in his Garden at Kensington Gore.

It was now ordered that the regular meetings of the Garden Committee be held in the months of April, June, August, and September. 1810

In accordance with a recommendation of the Committee, it was ordered by the Court of Assistants, that each Member of the Society pay in future ten shillings and sixpence annually towards the support of the Garden, instead of six shillings as heretofore. The former sum is still (1878) paid annually by every Member of the Society. 1811

In the month of December died Mr. John Fairbairn, who had held the office of Gardener upwards of thirty years. His great age and consequent infirmities had occasioned some neglect in the management of the Garden during the latter years of his life, and the consequences were very detrimental to the well-being of the establishment. The Court of Assistants determined to be influenced in the appointment of a successor solely by the merits of the respective candidates. 1814

Sir Joseph Banks, and Sir James Edward Smith, having honoured the Court with letters, recommending Mr. William Anderson, at that time principal gardener to James Vere, Esq., as a person eminently qualified to fill the office of Gardener, he was elected accordingly, with a salary of 100*l.* per annum and the usual apartments for his residence. At the same time, a pension of 40*l.* per annum was given to the widow of the late Gardener, with permission, by Mr. Anderson's consent, of her continuing in her apartments, but she did not live long to enjoy these benefits.

1815 The Garden was now in such a state as to require immediate attention, both from the great degree of dilapidation which had taken place in the different buildings, and from the long-continued neglect of cultivation. The Court of Assistants being determined to maintain the establishment with energy, prepared to adopt such measures as appeared to them best calculated to restore its former state of eminence, and even to advance it to as high a rank in the scale of exotic gardening as the improved state of that art would require.

Mr. Anderson, the Gardener, having been desired to give his opinion as to the best methods to be adopted for this purpose, recommended the following alterations; namely, that the old building immediately behind Sir Hans Sloane's statue, which had long ceased to be useful even as a tool-house, be pulled down, and the materials of it employed in erecting a room behind the little green-house, for the under-gardeners to sleep in, in which situation the apartment would be particularly useful to them, being near the flues, which, in the winter season, require the constant attention of one man during the night. That the flues in the several houses in the Garden, being badly constructed, be rebuilt on an improved principle, which would cause a considerable saving in the article of coal. That a pump be fixed in the Garden for a supply of Thames water, spring water being found to be injurious to plants. That the internal Garden formerly employed for culinary uses be appropriated to the purposes of Instructive Horticulture, a branch of Botanical Science which may be made very useful to society.

These recommendations obtained the approbation of the Committee, and, together with a general repair of the buildings, were directed to be carried into immediate effect.

The funds of the Society, however, being considered inadequate to accomplish these extensive designs in a suitable manner, a subscription among the Members was again proposed, and with the usual readiness was agreed to, according to the subjoined scale of payment; namely,

	£	s.	d.
Members of the Court of Assistants, each . .	3	3	0
,, of the Livery	2	2	0
,, of the Yeomanry	1	1	0

The total sum raised by these Subscriptions was 494*l.* 2*s.*

About this time, the Horticultural Society of London, being desirous to obtain and naturalise, as far as the climate would permit, all the cultivated varieties of vegetables employed as human food and as condiments, were anxious to possess the use of a garden to enable them to raise and increase all such rare and improved esculent plants and fruits as their extensive intercourse promised to supply, and which they proposed should be gratuitously distributed.

They therefore made the following proposal to the Court of Assistants; namely, that they might be allowed to use a part of the Physic Garden for the above purpose, under the entire government of the Court of Assistants and under the direction of their principal Gardener. They were ready to pay all the expenses attending the plan, together with the charge of an assistant Gardener.

The Committee, having maturely considered this proposal, declined accepting it on the terms above mentioned, but at the same time they expressed their readiness to direct any experiments to be made by their Gardener, in the science of Horticultural Gardening, upon application being made to them for that purpose. A part of the Garden, near the river, was consequently employed in promoting this design, which was carried out by planting various fruit-trees and esculent vegetables.

In consequence of a representation made to the Lord Mayor, by the Clerk of the Society, as to various depredations made on the shore before the wall of the Garden, the Water Bailiff examined into the complaint, and engaged that no further encroachment should take place.

1816 The superintendence and direction of the great improvements and repairs which had lately been carried out, having devolved chiefly on William Simons, Esq., Upper Warden, the thanks of the Committee were given to him for his judicious and economical application of the moneys so liberally contributed by the Members of the Society, in the construction of the different hot-houses, flues, &c., in the Garden. The total amount of money expended in these repairs, alterations, and new erections, was 574*l*. 13*s*. 10*d*.

1817 The ordinary expenditure of the Garden this year was, 535*l*. 8*s*. 11*d*.

1818 The Barge-house lately occupied by the

Society for their own use, being no longer required for that purpose, was let on lease to Mr. Lyall, of the Swan Brewhouse, for twenty-one years, from Michaelmas 1818, at a rent of 10*l.* per annum. That gentleman was permitted to open a window overlooking the Society's Garden, on paying annually 5*s.* for this privilege, and signing an agreement to stop it up, whenever required to do so, by a previous notice of three months. It appeared afterwards that the wall of this Barge-house was in so ruinous a state, in consequence of a common sewer running close to, and forming a part of it, that considerable repairs were necessary to be done to the wall. These repairs were therefore executed at the expense of the Society.

The Commissioners of Sewers having been applied to, and being requested to direct the course of the sewer to be altered in order to avoid future injury to the building, they gave directions accordingly. The expenditure of the Garden this year was 505*l.* 5*s.* 3*d.*, according to the subjoined statement.

	£	*s.*	*d.*
Labourers' Wages . . .	130	15	6
Gardener's Salary . . .	100	0	0
Sundry Bills	158	10	4
Ground Rent	5	0	0
Taxes	63	6	2
Committee Expenses . . .	47	13	3

Writing in the year 1820, Mr. Field, in reference to the tenure on which the Garden is held by the Society of Apothecaries, states that by the deed of conveyance from Sir Hans Sloane, two thousand plants were required to be presented to the Royal Society (see p. 37), but that this condition had been long before fulfilled. A much larger number, he proceeds to state, had been given than the condition demanded, but it was not easy to ascertain when the presentation ceased.

1820

By an extract from the minutes of the Royal Society, furnished by William Thomas Brande, Esq.,[1] one of their Secretaries, it would appear that the last presentation of plants took place on the 17th February, 1774, being the 51st annual presentation, the whole amounting in all to 2550 plants. It is perfectly certain that the plants were presented long subsequently to that time, but the delivery must either have taken place at irregular periods, or if otherwise, the minute books of the Society of Apothecaries have not regularly noticed it. The last

[1] Professor Brande, D.C.L., F.R.S., the distinguished chemist, the friend and colleague of Sir Humphry Davy, and predecessor and afterwards colleague of Faraday; a Member of the Society of Apothecaries, of which he held in due time the office of Master, and in which he occupied for a long series of years the Professorship of Chemistry and Materia Medica.

presentation of fifty plants, mentioned in those minutes, is in October 1794, the last preceding that being in October 1791. The entries of former years appear to have been equally irregular.

The former respected Editor of these Memoirs, Mr. Henry Field, in drawing his work to a conclusion in the year 1820, offers a general view of the state and condition of the GARDEN at that period. He comments upon the many difficulties in its management which the Society of Apothecaries had been from time to time called upon to surmount, and the repeated struggles made in various ways to maintain even the very existence of the Garden; and he expresses his gratification at the eminent position which the establishment at Chelsea then occupied. It had within a few years revived from circumstances of great depression, and indeed from a state almost bordering on dissolution, and had at last attained a degree of reputation as a scientific undertaking which the warmest aspirations of its most cordial supporters could have scarcely anticipated. He attributes, and with justice, its flourishing condition to the steady support it had always received from the Society of Apothecaries, and he bestows a warm tribute of praise on the skilful management of the Gardener, Mr. Anderson.

A considerable number of new plants, he says, have been introduced, and many thousands of seeds sown in the Garden within the few years since Mr. Anderson's appointment. Various improvements suggested by him have been already mentioned, and, in addition to these, the employment of sand as a medium of conveying warmth to the plants, instead of tan bark, had given promise of a permanent improvement, as a means of producing a more equable diffusion of heat, and at the same time causing a diminution of expense, both of which are important desiderata in exotic gardening.

In previous pages due credit has been accorded to the merits of the officials, who, in the respective capacities of Manager, or of Præfectus Horti and Botanical Demonstrator, or as Gardener, up to the period at which we have now arrived, had presided over the cultivation of the Plants, or promoted the advancement of Botanical knowledge; but it may not be superfluous to repeat that the names of James Petiver, James Sherard, Isaac Rand, Philip Miller, Joseph Miller, William Hudson, William Curtis, and William Forsyth, will always be memorable no less in the annals of horticultural and botanical science in general, than in the history of the rise and progress of the Botanic Garden at Chelsea. Nearly all these gentlemen are immortalised in the scientific world by the plants which bear their names, and the genera, *Petiveria*, *Sherardia*, *Randia*, *Milleria*, *Hudsonia*, *Curtisia*, and *Forsythia* will perpetuate to future generations the memory of the men who in various capacities have directed the progress of this Garden.

To another distinguished Botanist, who now appears as a veteran actor on the scene, Mr. Thomas Wheeler, the former Editor offers his well-merited tribute of admiration and respect; and to Mr. Wheeler and to others who have succeeded him as Officers of the Garden, the ensuing pages will in due course refer, to such an extent as the limited space of the present work will admit, and in such language as published or private records, aided in many respects by personal recollection, may supply.

In the year 1820, Mr. Wheeler had already filled for upwards of forty years the offices of Præfectus Horti and Botanical Demonstrator, and Mr. Field truly observes that in knowledge of practical Botany, and in the kind and familiar manner in which he communicated information to the students of the Society, he had been excelled by none of his predecessors, and had probably been equalled by few. For the truth of this eulogy Mr. Field thinks it only necessary to appeal to a great proportion of the then existing members of the Society who had been Mr. Wheeler's pupils, many of whom were good botanists, and who readily acknowledged the advantages they had gained from his instruction. Mr. Wheeler lived for many years after these passages were written, and there are many members living in the present day who can amply vouch for their justice and truth.

The plan of instruction in Botany in the Chelsea Garden at the period now referred to, was for the Demonstrator to discourse in the open air, at certain fixed periods during the summer months, to the members of the Society and their apprentices. The Medical plants were then, as they are now, arranged in systematic order in a certain part of the Garden, and the Demonstrator showed them to his pupils, explaining, as he went from bed to bed, their botanical characters, their place in the Linnæan classification, and their uses in Medicine. Most of the species in question are still collected together in the same part of the Garden, although the Linnæan classification has been replaced by the Natural Orders. At the period when these Memoirs first appeared (1820) the apprentices were allowed tea in the afternoons after the Demonstrations, which were held once a month.

In reference to a subject intimately connected with the history of the Garden and its Demonstrators and Professors, and with the efforts made by the Society of Apothecaries to promote the knowledge of Botany, the present opportunity is an appropriate one for a brief notice of the HERBORIZING EXCURSIONS which for many years were conducted in the vicinity of London for the purpose of collecting and demonstrating indigenous plants. It is almost needless to remark that the physical conditions of the metropolitan suburbs are now so entirely changed, that similar excursions in the present day would be useless or unprofitable; while the facilities afforded by the establishment of railways have supplied a ready method of conveying naturalists to rural districts formerly in a great measure inaccessible to the majority of the inhabitants of London. But up to and for some years after the period now referred to, the suburbs of London were by no means destitute of picturesque scenery and were rich in indigenous vegetation, and no better region could be selected, especially for the instruction of the botanical tyro, than the belt of country which surrounds the dwellings of our metropolitan population.

The first institution of Herborizing walks by the Society of Apothecaries was in the year 1633, at which time only one such walk was appointed to take place in each year, although other excursions were made by individual members of the Society for their own instruction. These walks seem to have been established for the benefit especially of the younger members of the Society and their apprentices. Thomas Johnson, the learned editor of the once-renowned "Gerard's Herbal," and a member of the Society, has left graphic accounts of some of these early excursions in some small volumes entitled "Iter in Agrum Cantianum" (1629), "Ericetum Hamstedianum" (1632), "Mercurius Botanicus, sive Plantarum gratiâ suscepti Itineris, anno MDCXXXIV, Descriptio," and a fourth entitled "Mercuri Botanici Pars Altera," bearing the date of 1641. These books are written in Latin, like most other scientific works of the period, the members of the Society who took part in the excursions calling themselves "Socii Itinerantes," and they appear to have been joined in their rambles by some of the Physicians of London and the provinces, and by others. Mr. Johnson, in his

Mercurius Botanicus, after alluding to his previous excursion in Kent, relates, in his first journey, his adventures and those of his companions in the south-western parts of England, travelling to Reading, and thence to Bath and Bristol, on to Southampton and the Isle of Wight, and home by Portsmouth and Guildford, the whole excursion occupying twelve days. The plants found on their way are enumerated, their Latin and English names being given, and the natural features of the country are briefly described. That Mr. Johnson and the other "Socii Itinerantes" were not insensible to the influence of hospitality is sufficiently clear from the gratitude expressed for kindness received in their wanderings, and that they did not despise the good things of the table is shown by the following quotation given by the author, from Virgil, when referring to his reception at a friend's house at Mangersfield.

> "Tum dapibus mensas onerant et pocula ponunt.
> Postquam prima quies epulis, mensæque remotæ,
> Crateras magnos statuunt et vina coronant."

Which, in Mr. Johnson's sense of the passage, may perhaps be rendered as follows:—

> "Spread by direction of the generous lord,
> A copious banquet loads the festive board,
> And when from solid fare the guests have ceased,
> With bumper cups they crown the jovial feast."

In the second excursion recorded by Johnson, the travellers explored the botany of Wales, their course being taken by Chester, and thence to Flint, Carnarvon, the Isle of Anglesea, Bangor, and home by Ludlow, Hereford, Gloucester, and Oxford. Many plants are described which did not appear in the first catalogue, some being collected under somewhat dangerous circumstances on the heights of Snowdon. The same pleasant style is observable in the description of the objects seen, the places visited, the persons met, and the hospitalities received. A rather amusing incident is recorded of a learned Doctor of Divinity, very fond of Botany, and who joined the other "Socii Itinerantes" at Chester, but who had been so ill-received at an inn at Stockport that, on departing in the morning, he left the following verses on the wall of his bedroom; but as the landlord probably knew nothing of Latin, the broad hint may never have reached him.

"Si mores cupias venustiores
Si lectum placidum, dapes salubres,
Si sumptum modicum, hospitem facetum,
Ancillam nitidam, impigrum ministrum,
Huc diverte, viator dolebis.
O Domina dignos, forma et fœtore ministros
Stock-portæ, si cui sordida grata, cubet."

Which lines may be thus freely rendered :—

"If, Traveller, you seek for quiet,
An easy couch, a wholesome diet,
A landlord with a smiling mien,
A chambermaid whose face is clean,
At Stockport you will never stay
But turn your steps another way;
But if in dirt and filth your soul delights,
At Stockport you may pass some pleasant nights."

A brief allusion has been made in a previous page (p. 48) to a botanical excursion made by James Sherard and Petiver. This took place in August, 1704, and the two botanists travelled in a chaise with two horses, and a servant rode a spare horse to change or put to as occasion offered. They went from London to Riverhead, thence to Sevenoaks, where they visited Dr. Fuller, author of the Pharmacopœia Extemporanea, "who entertained us," says Mr. Petiver, "with Venison pasty and the strongest drink in the county." At Tunbridge Wells "after botanizing all day, in the evening we visited the Wells incog., then refreshed with a bottle of pretty good Madeira and a couple of large Rabbits well dressed. Next morning we visited some friends, and for an hour showed our plants publickly, dined privately about eleven, and then left to encounter a rugged road (by Ticehurst and Wadhurst) which neither coach nor chaise had visited, it is said, in twenty years before. From such narrow ways Libera nos Domine, although, God be thanked, after many jostles and one small tumble we got safe in the evening to Brede." The account further says, "we had no opportunity of simpling" (gathering plants) "by the way, being in haste to reach our lodgings." Thence they went to Hastings, Rye, Lydd, New Romney, Sandgate Castle, Folkstone ("a base rugged town inhabited chiefly by fishermen"), Dover, Waldeshare, Knowlton, Deal, Sandwich, Isle of Thanet, Canterbury, Faversham, the Marshes near Sheppey, Rochester, and North-fleet, and so back to London. The account gives the names of the principal plants met with and the places where each plant was found. "We were in expectation of meeting *Orchis flo. conglomerato*, but the season was so far spent, and doubtless had we made this Voyage a Month or two sooner we should have met with more variety of rare and curious plants .... On the Banks of the Deep way near the Gravel Pitts on Blackheath we observed *Hyacinthus autumnalis* in seed and the flower all gone." (Sloane MSS. 3340.)

The following letter from Mr. James Petiver to Dr. Dale (see p. 63), announces an intended herborizing excursion in company with James Sherard :—

"SIR,

"Mr. James Sherard and I designing a Botanical Itinerary thro' some of the Hundreds, and in hopes of y^r good company and information, he desires me to acquaint you that we intend God willing to be at Braintree on Tuesday next before noon for ye remaining part of ye day, when we shall steer our course in our Journey further, we designing for Suffolk, Norfolk, &c. I also give myself no small pleasure in expectation of seeing Mr. Catesby's late acquisitions from Jamaica, and wch I hope I shall be able to give you some light into, having just printed something you have not yet seen relating to y^e Trees of those parts, which I shall bring with me, who am

"S^r y^r most humble $Serv^t$,

"J. P."

It would appear that the excursions of Mr. Johnson, the details of which include the earliest local botanical catalogues published in England, gave an impulse to similar excursions for educational purposes, not only in England but in other parts of Europe, and by the most distinguished Professors, as by Bernard de Jussieu at Paris, by the celebrated Haller at Göttingen, and by Strumph at Halle, and by Linnæus himself in the vicinity of Upsala. The last-named great naturalist writes a paper on the subject in his "Amœnitates Academicæ," founded on the botanical surveys he made with his students, and which were about eight in number every year. He remarks on the obvious utility of these excursions, which afford to students opportunities of seeing plants in their native places of growth, and also of making additions to the indigenous species already ascertained.

Dr. Pulteney, the learned Editor of the "General View of the Writings of Linnæus," refers to Thomas Johnson as the first promoter of what he calls "this laudable custom," and he alludes to the fact that the London Society of Apothecaries, after the endowment of Chelsea Garden, placed their botanical excursions under due regulations and fixed the periods for the "herborizations," and he particularly specifies the circumstances under which these excursions were made, as will be described in a future page.

The herborizations instituted by Linnæus at Upsala seem to have been conducted under circumstances peculiarly advantageous to those who attended them, and a considerable degree of *éclat* attended the proceedings. In Linnæus's Diary we are informed that that distinguished Professor, during his summer lectures, took out with him about 200 pupils, who collected plants and insects, made observations, shot birds, kept minutes, and, after having botanised from seven o'clock in the morning until nine in the evening, every Wednesday and Saturday, returned with flowers in their hats, and accompanied their leader with drums and trumpets through the city to the garden; and several foreigners and people of distinction from Stockholm used to attend these excursions. The drums and trumpets were used on the rambles for the purpose of calling the students together, and the evening procession was a kind of triumphal one, the excursionists being loaded with natural productions collected during the day.

"It is much to be wished," remarks Dr. Pulteney, "that a recreation in all respects so rational were established in other similar places of education, as it would at least diminish the number of votaries to other amusements which involve intemperance and prodigality."

These Botanical excursions are still maintained in almost all schools where Natural Science is taught, and they constitute an admirable method of making the student acquainted with the habits, the mode of growth, and the aspects of plants; and on the principle that what is *seen* makes a greater impression than what is merely described, the plan is better adapted than any other mode of instruction to teach the practical details of the Science of Botany.

> "Segniùs irritant animos demissa per aures
> Quam quæ sunt oculis subjecta fidelibus."
>
> *Horace.*

The herborizing excursions of the Society of Apothecaries were continued for the long period of two hundred years, but instead of one, as at first, the number was increased to six every year. Five of them were attended by the Apprentices of the Society, under the superintendence of the Demonstrator; and one of them, called the General Herborizing, was confined to the members only.[1] Of the former, two were held in the month of May, one in June, one in July, and one in August, and the walks were limited to the vicinity of the metropolis; the latter excursion was not so limited, but was extended, at the discretion of the Demonstrator, to the distance of thirty to forty or fifty miles, and generally occupied two or more days.

The object of this longer excursion was to collect a large number of indigenous plants, including those which grow near the sea, and the districts for collecting them were varied in different years. The plants so collected were exhibited on a fixed day in July to the Members of the Society, and some chosen guests, who consisted of the Office-bearers of the Royal Colleges of Physicians and Surgeons, and other

[1] The duties of the Demonstrator at these herborizings are set forth at p. 97.

distinguished visitors; and a general description of the nature and properties of the specimens was given by the Botanical Demonstrator, in whose duties this annual discourse was included. In former years the annual exhibition of Plants was held in some suburban hostelry, as at Blackheath, Putney, Hackney, &c., and after it had taken place the assembled company dined together at the expense of Stewards, who were chosen from the Members in rotation.[1] At these dinners haunches of venison formed conspicuous elements of the entertainment. In recent years, although the Annual Dinner has been continued, the Botanical Demonstration has been omitted, a circumstance very much to be regretted by all cultivators of

[1] In Petiver's "Adversaria," among the Sloane MSS. 3340, is a letter inviting Dr. Sloane (Sir Hans Sloane) to one of these banquets at the beginning of the eighteenth century,—

"WORTHY SIR,—

"The Master and Wardens having appointed me Steward for ye next Herborizing, and Putney Heath being a place not this year visited, I have determined ye Bowling Green there for ye lecture of Botany, where we shall dine on Tuesday next, and if your leisure will permitt I shall take it as a great favour to be honoured with y^{r} Presence and will add it to the many others you have most worthy Sir, been pleased to confer on

"Y^{r} most obedient and humble Servt,

"J. P."

Botanical Science, especially in its connexion, long insisted upon, and still acknowledged, with the study and practice of Medicine.

The other herborizing excursions, five in number in each summer, were open to the Apprentices of every Member of the Apothecaries' Society; and many are still living, to whom these summer rambles afford the most agreeable reminiscences. Days spent in healthy walks in the open country, in the company of youthful friends, and under the superintendence of kind and intelligent guides, must under any circumstances be regarded as affording one of the best and purest kinds of human enjoyment; but, when to the ordinary interest of these rambles is added a love for the beautiful productions of Nature which spring unbidden from the soil, the enjoyment is enhanced a hundredfold. As time rolls on in such pursuits, and as a taste for Botany is more and more developed, the aspect of the country becomes every day more familiar, and every tree and flower is enrolled as it were among the number of personal acquaintances, and the discovery of some new or unusual form of vegetation affords an agreeable variety to the pleasure of former acquisitions.

The custom was for the Apprentices to assemble punctually at six o'clock in the morning at St. Bartholomew's Hospital, while Mr. Thomas Wheeler held the post of Apothecary at that institution. But after his resignation of the Demonstrator's post, and when his son had succeeded him in the office, the youths met at Apothecaries' Hall, where they were received by the Botanical Demonstrator. An official of the Society carried a large metal box for the collection of the specimens, and each student carried a similar though smaller box, slung round his shoulder, for the same purpose. None carried umbrellas or great-coats, such incumbrances being considered useless and superfluous, and if the rain came down, the clothes were allowed to dry when the sun shone again.

The first appearance of Mr. Thomas Wheeler was certainly striking. A short, wiry, and thin old man (for at the time to which these reminiscences refer he was between seventy and eighty years of age) he entered with the alacrity of youth upon the scene, with an old hat in one hand, and a botanical knife in the other, with a pair of massive spectacles covering his grey and keen eyes, and clad in an old threadbare black coat and waistcoat and breeches, and a pair of long leather gaiters. But those who might be inclined to smile at his somewhat *outré* appearance were soon convinced that they were in the presence of no common person, and that the rough outer husk covered as true and genuine a man as ever adorned the profession of Medicine, or by his scientific and literary attainments shed lustre upon the Society of Apothecaries. More of this gentleman will be said hereafter, but it is impossible, even at the risk of incurring the charge of repetition, to dissociate his name from a record of the Society's Herborizings, which he superintended and personally conducted for the long space of fifty-five years.

It is true that, for the last thirteen years of that period, his son, Mr. James Lowe Wheeler, occupied the position of Botanical Demonstrator; but the veteran always accompanied the excursions to the last, was the prominent figure in the procession, was the guiding star of the botanical party, and excelled all the rest in the brightness of his intelligence, the extent of his information, and the activity of his movements. Looking back at the period now referred to, and recollecting the extent of the walks which are about to be described, it is really wonderful how this octogenarian preserved his animal spirits throughout those long yet delightful summer days, and how his physical energies enabled him to overcome the fatigue which might have wearied many a younger and more robust man.

He was an excellent, and indeed for his period, a profound Botanist, and withal a classical scholar, and he conveyed his information most readily in all departments of learning; and probably many of his peripatetic discourses may have left beneficial remembrances in the minds of his hearers on other matters besides the knowledge of indigenous plants. He was very particular about what is called the *quantity* of Greek and Latin words, and he seemed to be, or perhaps really was, horrified at hearing any pupil make a mistake in this respect. If any of them, for instance, unluckily pronounced *Anemŏne* for *Anemōne*, or *Arbōres* for *Arbŏres*, he would suddenly stop, and, summoning all the students, would shout with a loud voice the proper pronunciation of the word, warning them to be particular in future, and laying stress upon the difference between a scholar and an ignoramus. On the other hand, he was delighted when he found any of the youths giving evidence of a sound classical education, and he would frequently halt on the way in order to deliver some moral axiom illustrated from the vast stores of his own extensive learning. This was all done, however, in such a humorous and good-natured manner that a journey on foot of twenty miles during the day was made attractive, and secured the attention of the pupils, who were at first amused by his eccentricities, but afterwards impressed by his varied stores of information. On the subject of Botany, and especially indigenous Botany, as has just been mentioned, his knowledge was profound, and on such difficult matters, for instance, as the distinctions of the Grasses, the Sedges, the Umbelliferæ, the Compositæ, the Rushes, he was never at fault, but he both gave to every specimen its right name and explained minutely the points on which each species differed from another. Those who have experienced the difficulty of acquiring correct information on such points will understand what must have been the value of Mr. Wheeler's instructions.

The Students, having thus assembled in the bright summer mornings, pursued different routes on different occasions. Sometimes the excursion was to the north-west, passing through Islington,[1] and what were then known

[1] It is amusing to find in an old botanical book (anterior, however, in date to the period now referred to) that the Deadly Nightshade (*Atropa belladonna*) is described as growing wild *in a ditch in Goswell Street on the road to Islington!*

as White Conduit Fields (now covered with houses) then to the Copenhagen Fields, now converted into streets; thence to Kentish Town, and through the fields (now fields no longer) to the lower part of Hampstead Heath, and on to "Jack Straw's Castle," which still preserves its name and its reputation; and there the company sat down to a homely but abundant breakfast of tea and rolls and butter. After this repast the Demonstrator and the pupils scattered themselves about the Heath, gathering the ferns and the heath-plants which are still to be found there although in diminished number, and thence extending their rambles round Finchley, or Hendon, or Caen Wood, and returning to the Castle to dinner, which consisted of substantial joints of meat and pudding, and a moderate allowance of table ale. (Formerly, the apprentices were once a year at least indulged in wine, for at a Committee held at Chelsea Garden in the year 1823, it was ordered "that the pupils at the private Herborizings should be allowed a bottle of wine among four, and a bottle of cyder between two, but that no porter or other malt liquor should be allowed except table beer.") Immediately after dinner the large metal box carried by the attendant was opened, and the plants collected during the excursion were produced and exhibited to the students, who were seated on each side of a long table. The name of each plant was given, together with the peculiarities of form or structure by which it was distinguished, and if it possessed any medicinal qualities it would receive special attention as to its therapeutical uses. Thus an hour or two were pleasantly and profitably spent, after which tea was provided and the company dispersed, the students finding their way home as best they could, the return journey being generally performed, like the rest of the excursion, on foot, for at that time railroads and omnibuses were not in existence Sometimes a spare seat might be obtained on the top of a stage-coach, and when the excursion was up or down the river, the youths would perhaps row themselves or be rowed home in a boat in the evening twilight.

The other excursions were to the south-east and the south-west, and always included the banks of the river, and thus a more varied collection of plants could be obtained. One route lay along the Kent Road and to Deptford, and on to Greenwich and to Blackheath, where breakfast was provided at the "Green Man," which has long been deprived of the rural character it once presented, the premises and grounds being now built over. The extensive heath, still fortunately preserved as one of the lungs of the metropolis, afforded an abundant supply of plants peculiar to such a locality, and the walk was continued to the banks of the river between Greenwich and Charlton, thence towards Shooter's Hill, and back by the heath to dinner at the "Green Man," with the roast joints, pudding, and table ale, the demonstration of plants, the tea, and the return journey.

The excursion to the south-west was directed from Blackfriars, along the south bank of the river by Westminster Bridge, Lambeth Palace, Vauxhall, and into Battersea Fields, then a rich field of indigenous vegetation, where such rarities in British Botany as the beautiful *Fritillaria meleagris*, the *Polygonum bistorta*, and the *Stratiotes aloides* would reward the search of the zealous student. Thence the morning walk was extended through Wandsworth, and the fields along the waterside, and on to Putney, where, at the "Star and Garter," then a noted house, and still retaining some of its ancient renown, the company was received by the genial host and regaled with the usual liberal supply of hot rolls and tea. After breakfast the ramble was directed along the towing-path between Putney and Hammersmith, and an abundance of plants were found both on the banks of the river and in the marshes and ditches lying parallel to the course of the stream. Diverging from the Thames at Hammersmith the party struck across the common to Wimbledon and towards Richmond, and thence back to Putney to dine, the *pièces de résistance* at this meal being always eel pies, then a favourite luxury at this and other waterside resorts, before the introduction of steamboats had disturbed the fishery and compelled the eel-pie houses to seek their supply of fish at Billingsgate or elsewhere, instead of in the river running alongside. The eel-fishing is still continued in the upper part of the Thames, but the supply is greatly diminished.

During the Demonstratorship of Mr. Thomas Wheeler a prize, consisting of a copy of Hudson's "Flora Anglica," was presented at the close of each season to the young man who had been most successful in discovering and describing the greatest number of plants.

In drawing to a conclusion these reminiscences of the "Herborizings," now a thing of the past, of the Apothecaries' Society, it must be remarked that although the abandonment of the excursions was inevitable from the altered condition of the environs of London, yet the discontinuance of such means of instruction in Botany must be a matter of regret. By the demonstration of plants growing in their natural habitations, whether as stately trees or as lowly herbs; on the heaths and the commons; in the hedges and the ditches; by the side of rivers and subsidiary streams; among the grass of meadows; in the shade of woods and forests; on the bosom or in the recesses of lakes and ponds; amidst fields of standing corn; on old walls or the trunks of trees; or even on heaps of rubbish;—by such means a far more accurate idea is conveyed to the mind of the richness and the inexhaustible variety of the Vegetable Kingdom than could be communicated by any mere verbal or written description however accurate, or by pictorial representations however faithful, while the benefits conferred by such researches on the bodily health are too obvious to be insisted upon. With regard, also, to the Flora surrounding the metropolis, that is to say, the vegetable productions growing wild within a distance of ten or twenty miles from the centre, it may be interesting to observe that specimens of at least one hundred Natural Orders may be found within this area, and the botanist whose sphere of inquiry is limited by local circum stances may still find abundant materials for his study at no great distance from the busy and crowded haunts in which he is perhaps compelled habitually to reside. The wild plants are indeed banished to a rather greater distance than formerly, but the railroads and the steamboats have bridged over the intervening space, and now afford abundant facilities for visits to the country.

The Editor of the present Memoirs has himself found wild specimens of the following Natural Orders within the area above mentioned, and probably other explorers might add to the list. The primary divisions adopted are those of Jussieu.

1. *Acotyledones.*
Filices
Equisetaceæ
Characeæ
Musci
Lichenes
Fungi
Algæ

---

2. *Monocotyledones.*
Alismaceæ
Butomaceæ
Hydrocharidaceæ
Amaryllidaceæ
Iridaceæ
Liliaceæ
Orchidaceæ
Juncaceæ
Dioscoreaceæ
Pistiaceæ
Orontiaceæ
Naiadaceæ
Typhaceæ
Araceæ
Cyperaceæ
Graminaceæ

3. *Dicotyledones.*
Ranunculaceæ
Papaveraceæ
Nymphæaceæ
Cruciferæ
Fumariaceæ
Umbelliferæ
Malvaceæ
Geraniaceæ
Tiliaceæ
Hypericaceæ
Saxifragaceæ
Onagraceæ

3. *Dicotyledones* (*continued*).	Violaceæ	Vacciniaceæ
	Droseraceæ	Campanulaceæ
Circæaceæ	Linaceæ	Plantaginaceæ
Haloragaceæ	Alsinaceæ	Plumbaginaceæ
Salicaceæ	Silenaceæ	Dipsacaceæ
Corylaceæ	Crassulaceæ	Valerianaceæ
Betulaceæ	Chenopodiaceæ	Cichoraceæ
Cistaceæ	Polygonaceæ	Cynaraceæ
Rosaceæ	Primulaceæ	Asteraceæ
Pomaceæ	Solanaceæ	Rubiaceæ
Amygdaleæ	Callitrichaceæ	Cinchonaceæ
Leguminosæ	Cucurbitaceæ	Caprifoliaceæ
Urticaceæ	Aristolochiaceæ	Gentianaceæ
Ulmaceæ	Scleranthaceæ	Convolvulaceæ
Euphorbiaceæ	Lythraceæ	Oleaceæ
Resedaceæ	Coniferæ	Orobanchaceæ
Rhamnaceæ	Rhinanthaceæ	Scrophulariaceæ
Aceraceæ	Apocynaceæ	Verbenaceæ
Oxalidaceæ	Ilicaceæ	Labiatæ
Polygalaceæ	Ericaceæ	Boraginaceæ

1820 At the close of this year Mr. Thomas Wheeler resigned the Demonstratorship of Botany, which he had held for the long period of forty-two years, and a cordial vote of thanks was presented to him by the Court of Assistants for his valuable services.

Mr. Thomas Wheeler, F.L.S.

The following memoir, the first we believe presented to the public, with the exception of a short and in some respects incorrect obituary notice in the Journal of the Linnæan Society, will be read with some interest by those who still remember this very remarkable man, and it may not be deemed out of place even by the general reader. Mr. Thomas Wheeler was born in the year 1754, and received his elementary education under Mr. Garrow, the father of the eminent lawyer, Sir William Garrow, and he was afterwards a pupil at St. Paul's School. At this institution, celebrated as it always was, and is still, as a first-rate classical school, Thomas Wheeler had to endure the very severe discipline on the part of the head-master which was formerly considered a necessary part of a boy's education. This severity, of which corporal chastisement formed one of the main features, has now been happily relaxed in all educational establishments, but it cannot be denied that under such rigorous treatment some of the most eminent scholars have been fostered and developed. Be this as it may, and whether in consequence, or in spite of the discipline to which he was subjected, Wheeler became an excellent classical scholar, and to a great proficiency in the Greek and Latin languages he added a knowledge of Hebrew.

After leaving St. Paul's he was apprenticed to the Messrs. Walker, of St. James's Street, then Apothecaries to the King and Queen, but he does not seem to have enjoyed this period of his life much more than he did that of his pupilage under the strict and severe head-master of St. Paul's. In those days the duties of an apprentice were irksome, and to a certain extent degrading, and one of the foibles of Mr. Wheeler's character was an impatience of control. After being emancipated from what he considered, whether rightly or wrongly, the "house of bondage," he studied Medicine at St. Thomas's Hospital, attending the anatomical lectures of Mr. Hewson, and the Chemical and Medical courses of Dr. Fordyce, from the latter of whom he received great personal kindness, which lasted through life. At an early period he evinced a great fondness for the study of Botany, in which he was much encouraged by William Hudson, Botanical Demonstrator to the Apothecaries' Society, to whom reference has already been made in these pages, and after the resignation of Mr. Hudson and his successor Mr. Curtis, he was appointed to the vacant office in 1778, being then only twenty-four years old, and he held it for forty-two years. In 1800 he was elected Apothecary to Christ's Hospital, and six years afterwards to the similar, but far more important and responsible post, at St. Bartholomew's Hospital, where he remained for fourteen years, discharging most creditably and faithfully his onerous duties in that great institution. In 1820, he resigned the Apothecaryship of St. Bartholomew's and also the Botanical Demonstratorship of the Apothecaries' Society, but, as has before been mentioned, he continued to accompany the Herborizing excursions until they were abandoned, at which time (1834) he was eighty-one years of age. After this latter period he resided alternately with one or other of his sons, and having enjoyed a peaceful and happy old age, he died without a struggle, and from a mere decay of nature, in his 94th year at the house of his eldest son, Mr. Thomas Wheeler. He had filled the usual honourable offices in the Apothecaries' Society, of which he became Master in due course of time, and he was moreover a member of the first Court of Examiners appointed by the Society under the Act of 1815. He had married early in life, but was left a widower at the age of forty, with six children. The following entry, relating to this event and taken from his family Bible, will convey some idea of his simple piety which, with other good qualities, formed a leading feature in his character :—" The light of my eyes and the joy of my heart, my dear wife, was this day taken from me. Lord, Thy will, not mine, be done." He did not marry a second time.

As a botanist, Mr. Wheeler's attainments were of a very high order, and he might be considered as one of the best botanists of his day. He was enthusiastically attached to the doctrines of Linnæus, the "novum sidus," as William Hudson designated him, which had then just illuminated the botanical world, and he, of course, taught and admired the Linnæan System, which in Mr. Wheeler's time was the only possible method of teaching the science of Botany to the rising generation; for the views of Jussieu were not propounded until 1789, and were not at first received with much favour, owing to the difficulty of learning the system of that distinguished Professor. As a teacher of Botany Mr. Wheeler was eminently successful, and he had the faculty of inspiring his pupils with his own enthusiasm for his favourite pursuit.

Many anecdotes, some of a humorous and others of a serious and touching character, are extant in reference to Mr. Wheeler, and some of these will now be offered to notice.

During the early part of his long life he was enrolled among the Volunteer Military Forces, and was ordered out during a period of public disaffection to quell a riot which was apprehended, and in reference to this circumstance he used to say, "If I had been ordered to fire upon the people, I should certainly have done so, offering up at the same time a prayer that my bullet might not take effect."

He was very tenacious of his professional

dignity, and once when a Reverend Prelate seemed to question the treatment of a patient under his care at the Hospital, and expressed his own opinion in a somewhat inflated manner, Wheeler replied by imitating in his answer the pompous and arrogant tone of the Bishop, whereupon a bystander said, "Why, Mr. Wheeler, what a proud man you are!" and he replied at once, "Inter superbos tantùm, superbus." So it was indeed with him, he was "proud, only among the proud," and to the poor he was the kindest of the kind. Once discussing with his wife the expediency of devoting some rather large portion of their limited means to the relief of a necessitous object, he said, "We are too poor, my dear, to curtail our charities."

The following anecdote illustrates his promptitude and energy when human life was at stake. On one occasion of the Herborizing excursions, when the route lay along the banks of the Thames, and he and the students were searching for botanical specimens, they witnessed the accidental upsetting of a boat on the river, the rowers being chiefly young lads. It was anxiously asked by all the party whether the boys were all saved, and the reply was, "Yes, we are all here;" but Mr. Wheeler would not be so easily satisfied, and seeing the boat floating keel uppermost, he shouted at the top of his voice, which was a very loud one, "Turn up the boat, for Heaven's sake, turn up the boat!" which was accordingly done, and a youth of fourteen was found under the thwarts insensible, whose life would have been lost but for Mr. Wheeler's determination and presence of mind.

Some other incidents were of a different and humorous character. It has been already mentioned that he was very peculiar in his style of dress, carrying simplicity in this respect almost to extremes, and a friend, once speaking to him on the subject, remarked in the words of Alexander the Great, "Ah, Diogenes, thy pride peeps out of every button-hole of thy coat!" A story is told that he was one evening sitting in a room at the back of the "shop," as the Surgery at St. Bartholomew's Hospital was called, with a number of Students who were bantering him and one another, and he was descanting on the folly of superfluities in dress, &c. Young Lawrence (afterwards the distinguished Sir William Lawrence, Bart.) said with his usual assumed gravity, "Well, but Mr. Wheeler, how can you support such a doctrine while you wear such a superfluity as this!" lifting up the small *queue* or pigtail which

Wheeler wore. Thus taken aback, the old man confessed that it was superfluous: "Yes, my dear sir, you are right, we are too prone to preach one thing and to practise another. I never thought of it; cut it off, sir, pray cut it off," and Lawrence forthwith performed the amputation requested.

Another story is in connexion with the Herborizings previously described. At one of these, called the "Grand Herborizing," which took place once a year, the Demonstrator accompanied by one or two other persons, was exploring some localities in Kent in search of plants for the annual lecture. On this occasion they were in the neighbourhood of Maidstone, and the party were driving in an open barouche. Mr. Hurlock, a well-known member of the Apothecaries' Society, was inside with James Lowe Wheeler and two others, and Mr. Wheeler was seated on the box by the side of the driver. Mr. Hurlock, who told the story, described the appearance of Mr. Wheeler (which all who knew the old gentleman will readily recognise) with his hat off, his thin light hair blowing about his face and his large spectacles on his nose, alternately laughing and chatting with the driver, and diving into his hat with his huge pocket-knife, separating and examining a bundle of wild plants. Such a figure naturally attracted attention along the road, and, when stopping at a turnpike-gate, the party were rather surprised by the evident interest and eagerness of the toll-keeper, as he scratched his head, and, pointing to Mr. Wheeler, exclaimed in his blunt Kentish dialect, "So ye ha got him at last!" This was incomprehensible to all the party until they arrived at a small inn close to the parish of Barming, where they read a placard offering a reward for the capture of an escaped lunatic!

From the age of forty to the period of his death at the age of 94, Mr. Wheeler entirely abstained from fermented liquors, but not from any ascetic feeling, as he never moralised on what is called teetotalism or objected to the moderate use of alcohol by other people. He often explained the reason of his singularity in this respect, which was merely that he had left off such beverages on the occasion of a very serious illness, and ever afterwards had found himself better without them. His appetite however was excellent, and at the breakfasts and dinners of the Herborizing excursions, the heartiness with which he attacked the plain but abundant fare always provided was not surpassed by the youngest student, with whom, as before remarked, his own walking powers might advantageously to himself be compared. In fact he deserved his breakfast and dinner quite as much as any of the juvenile party.

It is stated in the brief obituary record to which a slight allusion has been made (p. 152), that "in manners and habits he was distinguished for child-like simplicity," and this is in a great measure true. He never jested with sacred things, and never uttered a joke which could raise a blush on the most modest cheek. His discourses, which, like those of Plato, were delivered as he walked, were not confined to the science of Botany, and he never lost an opportunity of saying a wise and instructive word to his young disciples, some of whom may even now confess that not a few of the principles which guided them in mature years were early instilled into their minds by the earnest teachings of this simple-hearted old botanist. "Maxima reverentia debetur pueris" he thought and practised, and he fully carried out the precept of Horace,—

> "Nil dictu fœdum visuve hæc limina tangat
> Intra quæ puer est."

To such teachings as Mr. Wheeler's those who remember them may apply the lines of Thackeray,—

"And if in time of sacred youth
We early learn to love and pray,
Pray Heaven that early love and truth
May never wholly pass away."

Charles Kingsley's ideal character of the perfect Naturalist in a remarkable manner describes Mr. Thomas Wheeler:—

"As one who should combine in himself the very essence of true chivalry, namely, self-devotion, whose moral character, like the true knight of old, must be gentle and courteous, brave and enterprising, and withal patient and undaunted in investigation, knowing (as Lord Bacon would have put it) that the kingdom of Nature, like the kingdom of Heaven, must be taken by violence, and that only to those who knock earnestly and long, does the Great Mother open the doors of her sanctuary. He must be of a reverent turn of mind too . . . . always reverent yet never superstitious, wondering at the commonest, yet not surprised at the most strange: free from the idols of size and sensuous loveliness . . . . holding every phenomenon worth the noting down; believing that every pebble holds a treasure, every bud a revelation; making it a point of conscience to pass over nothing through laziness or hastiness, lest the vision once offered and despised should be withdrawn, and looking at every object as if he were never to behold it more . . . .

"He must have that solemn and scrupulous reverence for truth, the habit of mind which regards each fact and discovery not as our own possession, but as the possession of its Creator independent of us, our needs, our tastes—it is the very essence of a naturalist's faculty, the very tenure of his existence; and without truthfulness, science would be as impossible now as chivalry was of old."—*Life of Charles Kingsley*, vol. i. pp. 406, 407.

An excellent portrait of Mr. Wheeler by

Henry Briggs, R.A., is placed in the great parlour of Apothecaries' Hall.

Among other pleasurable feelings connected with the preparation of the present Memoirs, one of the most agreeable to the Editor is the opportunity offered him for paying the above imperfect tribute to the memory of one of his earliest and best instructors.

Mr. Wheeler had six sons, all of whom were brought up to his own profession. He loved them all intensely, and his love was as fondly returned. They were Mr. Thomas Lowe Wheeler, for several years Chairman of the Court of Examiners, Dr. Charles West Wheeler, Mr. James Lowe Wheeler, Mr. William Lowe Wheeler, Mr. Lowe Wheeler, and Mr. Joseph Wheeler.

All of these, most of whom filled the usual honourable positions in the Apothecaries' Society, having led blameless lives, and each attaining to a ripe though not extreme old age, are now dead, and not many descendants remain. Among the few survivors are Mr. Thomas Rivington Wheeler, F.R.C.S. who worthily occupies the position of Secretary to the present Court of Examiners, and Mrs. T. R. Wheeler, daughter of Dr. Charles W. Wheeler, who, with her husband, inherits the literary and scientific tastes of their grandfather.

1821 On the resignation of Mr. Thomas Wheeler, the post of Demonstrator was filled by the appointment of that gentleman's third son, Mr. James Lowe Wheeler, who commenced his duties in this year. He followed very closely in the footsteps of his father, who, as has just been mentioned, accompanied him in all the Herborizing excursions and attended his other demonstrations and lectures. Mr. J. L. Wheeler, like all his brothers, was a thoroughly well educated man and a classical scholar, and though not so enthusiastically devoted to Botany as his father, he was not only very fairly acquainted with the principles and practice of that science, but he was also an excellent chemist, the science of Chemistry having begun to dawn upon the world at the period to which these Memoirs have now reached. Just as Linnæus in the middle of the last century had converted Botany from a chaotic assemblage of facts into a definite science, so at the beginning of the present century had Sir Humphry Davy performed a similar service for Chemistry; and the discovery of the metallic bases of the alkalies and earths, and the brilliant applications of Galvanism and Electricity to the synthesis and analysis of bodies, laid the foundation of the extensive mass of knowledge which, spreading its domain over all creation, "from the heavens above" (for the composition of the sun, moon, and stars is now known) "to the earth beneath, and the waters under the earth," constitutes the modern science of Chemistry.

Mr. J. L. Wheeler for many years lectured on Chemistry and Materia Medica at a private Medical school near St. Bartholomew's Hospital, and was there associated with his brother Mr. Lowe Wheeler, who lectured on Anatomy.

For some years after the appointment of Mr. J. L. Wheeler as Demonstrator of Botany, no very striking event is recorded in reference to the history of the Garden, and it must be taken for granted that the Demonstrations at the Garden, and the breakfasts or "teas" of the Apprentices, and the Herborizing excursions, were regularly held, and that the Garden itself was kept up in a state of efficiency.

It may perhaps, however, be interesting to present an abstract of the yearly expenses of the establishment for fifteen years, as recorded in the Minutes of the Garden Committee, contrasted with the funds available for the purpose, by which account it will be seen that the Society were compelled to meet a heavy deficiency every year. The ordinary sources from which the Garden is maintained are set forth in former pages of these Memoirs, and consist of fixed annual payments (called Quarterage) from each Member of the Society, of fines upon admission to certain offices of the Society, of proportion of fees from Apprentices, &c. The deficiency, therefore, represents the difference between these ordinary sources of income and the actual expenses of the Garden, consisting of Salaries to Demonstrator, Gardener, labourers, and taxes, coals, repairs, &c.

		£	s.	d.		£	s.	d.
In 1820	the expenses were	463	19	11	deficiency	211	16	5
1821	"	529	5	2	"	273	18	2
1823 [1]	"	555	5	3	"	284	7	9
1824	"	511	6	10	"	285	7	4
1825	"	662	5	9	"	419	18	3
1826	"	590	11	1	"	340	3	1
1827	"	692	18	6	"	456	18	6
1828	"	590	18	7	"	363	6	7
1829	"	753	5	2	"	518	5	2
1830	"	613	16	11	"	374	10	5
1831	"	893	5	10	"	640	6	4
1832	"	721	16	10	"	469	14	10
1833	"	688	8	2	"	444	11	8
1834	"	771	6	0	"	504	2	0
1835	"	735	11	6	"	470	5	0

The deficiency was supplied, as it still is, from the corporate funds of the Society.

[1] By some omission, not accounted for, the receipts and expenditure for 1822 do not appear in the Minutes.

Before the election of a successor to Mr. Thomas Wheeler in the post of Demonstrator to the Garden, a Committee was appointed to consider the duties of that office and to report upon the subject to the Court of Assistants. This Report was accordingly made in January 1821, and the recommendations made were founded essentially on the former minute relating to the same subject in the year 1773, as recited at page 97. Some alterations and additions, however, were suggested in order to render the office of Demonstrator as beneficial to the Members of the Society and their apprentices as the advanced state of science now required. The chief suggestions were that the Demonstrations in the Garden should be held on the last Wednesday in the summer months at nine o'clock in the morning; and that the Demonstrator was expected on each such occasion, or on other occasions if so directed, to employ some time "in explaining to the Students the systems of Botany, both Sexual and Natural, as taught by Linnæus and Jussieu; together with the principles of vegetable life, and the Structure, Physiology, and Medical Virtues of Plants, their Natural Climate, the alterations produced by culture, and the parts of them employed whether medicinally or as food for

1821

man and other animals." The other suggestions are on minor points, such as the direction of the Herborizing excursions, the Superintendence of the Apprentices, the delivery of the annual Lecture to the Members of the Society in July, the attendance at the meetings of the Garden Committee, and some other similar matters.

This Report was approved and confirmed by the Court of Assistants, and the Salary of the Demonstrator and Professor of Botany (as he was now called for the first time) was fixed at 50*l*. per annum. At a subsequent meeting of the Court of Assistants, in March of the same year, Mr. James Lowe Wheeler was elected to the vacant office.

The language quoted in the above Report of the Garden Committee proves that the Managers of the Garden were aware of the advances now being made in the Science of Botany, and that they wished their recommendations to be in keeping with the spirit of the times. It shows, for the first time in the history of the Garden, that the Natural System of arrangement was beginning to make itself known and understood; and it may be necessary briefly to allude to the principles on which that System is founded.

It has already been mentioned that very great difficulties exist in every attempt to classify the objects of the Vegetable Kingdom, and indeed for a great number of years any plan of systematic arrangement of plants presented only a hopeless enigma. Linnæus, however, happily hit upon a character which distinguished plants into two great divisions, according as they did or did not possess Stamens and Pistils, or, in other words, as they were Flowering or Flowerless Plants (*Phœnogamic* and *Cryptogamic*). He further subdivided the Flowering Plants into twenty-three Classes, and these again were subdivided into a number of Orders. This arrangement, founded mainly on certain arrangements of the Stamens and Pistils, and which has been briefly explained at p. 59, was avowedly artificial, but it was founded upon Nature, for Linnæus proved to the satisfaction of the scientific world that every Flowering Plant, without exception, possessed Stamens and Pistils, either or both, which are as essential to the continuance of the species as the sexes are in animals. The error, if error it were, in the System of Linnæus was, that this arrangement was founded exclusively upon the peculiarities of a single set of organs, and neglected the rest, such as the stem, the leaves, the fruit, and the seeds. Linnæus, however, was fully aware of the imperfection of his own system, and he himself proposed a Natural one, which he was unable to complete during his lifetime, partly in consequence of his imperfect resources, residing as he did in a small University town in Sweden, and partly from the inherent complexity and difficulty of the task.

But almost contemporaneously with the labours and the discoveries of Linnæus, a distinguished family of Botanists in France were labouring to establish what was, and is still known, as the NATURAL SYSTEM, which aimed at founding a Classification, not on the peculiarities of one set of organs, but on the general arrangement of the whole, and which should associate the configuration and structure of plants with their medicinal, dietetic, economic or other similar properties.

The Jussieus.

The family here referred to is that of the Jussieus, who were for nearly a hundred and fifty years distinguished in the annals of Botany, and one of whose descendants held till lately the Professorship of that science in the Jardin des Plantes in Paris. Antoine de Jussieu was born in 1686, was a physician in Paris, and was Professor of Botany in the Jardin du Roi, founded by Louis XV. He published various papers on Botany, and died in 1758. Bernard de Jussieu, his younger brother, was Demonstrator of Plants at the same Garden. He was born in 1698, and died in 1777. He also wrote various papers on Botany, and published an arrangement of the plants growing in the Garden of the Trianon at Versailles. Joseph de Jussieu, a third brother, was born in 1704, and died in 1779. He was sent to South America by Louis XV., remained there for thirty-six years, and made many discoveries and brought home many new plants. Antoine Laurent de Jussieu, the son of Bernard, was born in 1748, became Professor of Botany at the Jardin du Roi (now called Jardin des Plantes), was member of the Institute and of nearly every learned body in Europe. In the *Genera Plantarum*, published by him in 1789, he developed, to a fuller extent than ever had been done before, the NATURAL SYSTEM of Botany, which had been sketched out by Tournefort and by Linnæus, and by the elder Jussieus. He attained the advanced age of eighty-nine, and died at Paris. The writer of the present Memoirs, when a student in that city, witnessed the funeral of this great Botanist, which was conducted with great pomp, the coffin being surmounted by the deceased Professor's cap and the Cross and Sword of the Legion of Honour, and the ceremony being attended by all the Professors of the Jardin des Plantes in full costume, and also by the Dean and the Professors of the Faculty of Medicine, together with the usual military escort of a Chevalier of the Legion of Honour. Antoine Laurent de Jussieu was succeeded in his Professorship of Botany at the Jardin des Plantes by his son Adrien de Jussieu, who died in 1853.

As the limits of these Memoirs forbad the introduction of a full description of the Linnæan System, the same reason exists for the omission of any extended reference to the System first clearly developed by Antoine Laurent de Jussieu. It may, however, be stated briefly that the main feature of the latter system was the division of plants into three sections, founded on the presence or absence of the seed-lobes (cotyledons) and the single or double nature of those organs. Hence the names of these sections were *Acotyledons*,[1] *Monocotyledons*,[1] and *Dicotyledons*.[1] Nothing can be more simple than this arrangement; but the classes, which were fifteen in number, were founded on characters

[1] *α*, without; *μονος*, single; *δις*, twice; and *κοτυληδων*, a seed-lobe.

far more complicated, and they were subdivided into a great number of *Orders*, arranged according to their general structure, and hence called *Natural Orders*. The *Genera Plantarum* of Jussieu is written in Latin, and in the Introduction the author very ably shows the imperfections of the Linnæan System of classification, but he is not so successful in proving the faultlessness of his own. Still his system displays great ingenuity and profound learning, and unquestionably forms the basis on which all subsequent Natural Systems of Botany have been established.

As demonstrating the great difficulty of establishing a comprehensive and unobjectionable Natural System of Botany, it is sufficient to remark that the System of Jussieu was soon materially modified, though not essentially changed, by another great botanist, Auguste Decandolle, who was born in 1778, passed many years in France, where he studied and taught, and finally became established at Geneva, as Professor of Botany in that city and Superintendent of the well-known Botanic Garden in the immediate vicinity. Decandolle was a man of indefatigable industry, and he set himself the Herculean labour of describing the whole of the Vegetable Kingdom according to the Natural

System, and he actually commenced a work entitled "Systema Naturale" in 1818. But he soon found that the task was beyond his powers, or indeed those of any other human being, and he accordingly contented himself with writing an abstract of the proposed greater work, and which he termed "Prodromus Systematis Naturalis Regni Vegetabilis, sive Enumeratio Contracta Ordinum Generum Specierumque Plantarum hucusque cognitarum, juxta methodi naturalis normas digesta." (Outline of the Natural System of the Vegetable Kingdom, or abbreviated Enumeration of the Orders, Genera, and Species of Plants hitherto known, arranged according to the rules of the natural method.) This *abbreviated* work was begun in 1824, and though extending to fifteen volumes was left unfinished at Decandolle's death, which took place in 1841. The work was continued by his son Alphonse Decandolle, and by his grandson Adrien Decandolle, both of whom are still living, but the task is not yet finished.

The System of Decandolle is a modification of that of Jussieu, and is an improvement upon it. Decandolle adopts Linnæus's name of *Cryptogamæ* as synonymous with the *Acotyledones* of Jussieu; and the *Monocotyledones* and the *Dicotyledones* of the latter are synonymous respectively with the *Endogenæ* and *Exogenæ* of Decandolle, which names are founded on the arrangement of the woody layers of the *stem*. The Exogenæ are divided into four sub-classes, according (1) as the stamens are inserted into the receptacle (*thalamus*), (2) as they are inserted into the calyx, (3) as they are attached to the corolla, and (4) as there are no petals in the flower; and these sub-classes are accordingly termed *Thalamifloræ*, *Calycifloræ*, *Corollifloræ*, and *Apetalæ*. The Endogenæ have no definite sub-classes, but the Cryptogamæ are divided into *Acrogens*, which have an upright stem, and *Thallogens*, in which there is no proper stem and the whole structure consists only of cells. The system of Decandolle, although very far from perfect or unobjectionable, is perhaps the best and simplest of the Natural Systems hitherto proposed, and is even now most generally adopted.

The above remarks are essential as explaining the principles of the Natural System of Botany, and as introductory to the proceedings which occurred in the history of the Chelsea Garden between 1821 and 1829.

1829 For several years after 1821, however, the affairs of Chelsea Garden were conducted in the usual manner, and no occurrence of any importance is recorded. But a feeling appears to have gradually arisen in the minds of the Garden Committee that the establishment at Chelsea was hardly fulfilling sufficiently the objects of its founders. The Demonstrations and the Herborizings were limited entirely to the members of the Apothecaries' Society and their Apprentices, who were, in proportion to the whole Profession, comparatively few in number, whereas the Licentiates who had received qualifications to practise from the Court of Examiners since the year 1815 were now very numerous, and the students preparing for examination were increasing in number every year. The science of Botany, also, had made enormous strides during the last and present century.

It was therefore resolved at a Garden Committee held in 1829 that, in the opinion of the Committee, the Botanical Garden might with propriety be made more useful to the Profession at large than it had hitherto been; and the Master and Wardens of the Society were requested to form a plan for carrying such intentions into effect.

The first effect of this movement was a recommendation that the Garden should be opened on Friday the 3rd of July of this year between nine and eleven, and on every succeeding Friday at the same time; and that admission should be given by tickets to all such Medical Students as were pupils to the established Professors and Tutors in the Metropolis in Medicine, Chemistry, Materia Medica, or Botany. It was only necessary in order to procure admission that the pupil should bring a letter of recommendation from his Teacher, stating that such pupil had been attentive to his studies, and was desirous of improving himself in Medical Botany.

These recommendations were adopted, and the Garden was accordingly thrown open to the Students of the Metropolitan Schools of Medicine, and the result was so successful during the first summer when the experiment was tried, that a further Report was made on the subject, with a view of still more facilitating the acquisition of Botanical Science by the Students of Medicine.

The Report now alluded to was as follows:—

TO THE MASTER, WARDENS, AND ASSISTANTS OF THE SOCIETY OF APOTHECARIES.

GENTLEMEN,—

Your Committee, appointed to take into consideration the best method of rendering the Society's Garden at Chelsea more generally useful to the student in medicine by increasing his opportunity of improvement in the science of Botany, particularly as connected with the Medical Profession, have much satisfaction in referring to a Minute of the Court of Assistants held on the 23rd day of June last, containing a plan for admitting into their Garden under certain limitations all such Medical students as are pupils to the established professors and tutors in the metropolis, whether in Medicine, Chemistry, Materia Medica, or Botany.

This design having been carried into effect during the last summer, was received by the several professors and lecturers in the most gratifying manner, and thankfully embraced by more than 100 pupils. In consequence of this, your Committee recommend that a similar plan upon a more enlarged scale be in future adopted; the particulars of which design are contained in the following regulations:—

That the Garden be open every Wednesday during the months of May, June, July, August, and September, from 9 o'clock in the morning until 12 at noon, and that admission be given to all such medical students as are pupils to the established professors and lecturers in the metropolis, whether in Medicine, Chemistry, Materia Medica, or Botany, and also to the Apprentices of the several Members of the Society.

That there be every week a demonstration of all the plants contained in the Materia Medica department of the Garden, and of such other plants as the Demonstrator may think proper. Such demonstration to commence at 10 o'clock punctually, and that after such demonstration is finished there be a lecture delivered by the Demonstrator in some part of the building attached to the Garden, upon one or more of the following subjects, so as to form during each summer season a regular Course of Botanic Study, namely,—

1. The different systems of Botany, both natural and artificial, particularly those of Linnæus and Jussieu.
2. The Structure and Growth of Plants.

3. The different parts of Plants with their description and uses in the process of Vegetation.
4. The natural and chemical analysis of vegetable matter.
5. The medicinal uses of the most important articles in the Materia Medica, with observations on the best modes of preparing them. These remarks may be made either at the lectures or at the demonstrations at the discretion of the lecturer.

That the conducting these demonstrations and lectures be committed to the Society's Demonstrator of Botany, and that the monthly lectures hitherto delivered by him at the Garden be discontinued, as merging in and more effectually provided for in the lectures now proposed to be adopted.

That in order to give encouragement to diligence and talent, there be an annual examination of such students as may think proper to become candidates for the prizes intended to be given on these occasions. The examinations to be upon some or all of the subjects stated in the foregoing series of lectures, as well as upon their skill in the nomenclature of plants. No person to be admitted a candidate who has not attended these lectures and demonstrations at least eighteen days in one summer, or thirty days in two succeeding summers, nor shall any prize be awarded unless his examination be performed to the complete satisfaction of the examiner or examiners for the time being.

To prevent partiality or undue preference, no public professor or lecturer whose pupils are admitted to the Garden can be appointed an examiner.

The apprentices to Members of the Society having an annual opportunity of being candidates for prizes upon the ancient establishment, cannot be admitted candidates on these occasions either during the period of their apprenticeship, or subsequently to the conclusion of it.

That two medals, the one being of gold of ten guineas value, and the other of silver or bronze, be annually awarded to the two candidates who shall have passed the best and second best examination in manner hereinbefore mentioned, but no medal to be given unless in the opinion of the examiner or examiners the candidate shall be deemed deserving of it.

That in consequence of the additional service occasioned by these lectures the salary of the Botanical Demonstrator be increased.

(Then follow some regulations as to the mode of admission, the reception of the students, &c.)

That these Resolutions or such parts of them as are necessary for the information of the medical students be printed and a copy sent to each member of the Society, and to the several hospitals and medical schools of the metropolis.

APOTHECARIES HALL,
12th December, 1829.

JOSEPH HURLOCK, *Master*.
W. R. MACDONALD, } *Wardens*.
JOHN HUNTER, }
HEN. FIELD.
JOHN NUSSEY.
WM. KING.

1830 These recommendations were approved, and the salary of the Demonstrator and Professor was raised to 80*l*. per annum. The new regulations came into operation in the ensuing summer, and the Lectures were regularly delivered every Wednesday morning by Mr. J. L. Wheeler. The number of Students who availed themselves of the new privilege was very great, and from every Medical School in London numbers of youths flocked to the Garden either on foot, or (what was a very favourite mode of conveyance) in rowing boats from Westminster, thus combining healthy recreation and exercise with intellectual improvement.

In this year Mr. J. L. Wheeler, the Professor of Botany, published a Catalogue of the Medicinal plants in the Garden. It is written in Latin and is entitled, "Catalogus Rationalis Plantarum Medicinalium, in horto Societatis Pharmaceuticæ Londinensis, apud vicum Chelsea, cultarum. Curante Jacobo L. Wheeler, Societatis Botanices Professore, Soc. Linn. Socio, et in Chemia et Materia Medica Prælectore." It was by far the best catalogue of the medicinal plants in the Chelsea Garden ever published, and is, even now, an excellent practical guide to Medical Botany. The plants described are those admitted into the Pharmacopœiæ of London, Edinburgh and Dublin (all now consolidated into one, the British Pharmacopœia of 1867), and the properties and doses of the medicinal products of each are briefly but distinctly specified. The arrangement followed is that of Linnæus, but a Synopsis is also given of the plants according to the system of Jussieu. In the first part of the Catalogue, which consists of the classification and description of the plants according to the Sexual System of Linnæus, the Natural Order of that great Botanist is also given to each plant, and then the Natural Order of Jussieu.

It is not generally known that many of the Natural Orders, now supposed to have been invented by Jussieu and Decandolle were really first established by Linnæus, though sometimes under slightly different names. Thus, the *Umbellatæ* of Linnæus are the *Umbelliferæ* of Jussieu; the *Cucurbitaceæ* are common to both, and so are the *Coniferæ;* the *Gramina* of Linnæus are the *Gramineæ* of Jussieu; the *Filices*, *Algæ*, and *Fungi* are names adopted by both. Again, the *Siliquosæ*, the *Verticillatæ*, the *Luridæ*, the *Stellatæ* of Linnæus correspond respectively to the *Cruciferæ*, the *Labiatæ*, the *Solaneæ*, and the *Rubiaceæ* of Jussieu, and a great number of similar instances might be adduced.

Towards the close of the year, when the Apothecaries' Society had done so much to promote the Science of Botany among Medical Students, Mr. Lindley, who had lately become eminent by his appointment as the first Professor of Botany at the then University of London (now University College), appropriately dedicated to the Court of Examiners the first edition of his "Introduction to the Natural System of Botany." In the Dedication he says, "Measures have lately been taken by the Society of

Apothecaries, which cannot fail to exercise a most beneficial influence upon Botany, and which must have been viewed with feelings of deep interest by all friends of the Science. As a humble individual, whose life is devoted to its investigation, I am anxious to take the present opportunity of expressing my sentiments upon the subject, by very respectfully offering for your acceptance a Work, which it is hoped will be found useful to the Student of Medical Botany."

The lectures at the Garden were now regularly delivered every Wednesday morning during the summer months, and the weekly visits of the Students became enrolled among the established institutions of the Medical curriculum. At the same time the "Herborizing" excursions were continued as usual, five for the Apprentices, and one, the "Grand Herborizing," with the subsequent Demonstration and Dinner, for the Members of the Society.

Three prizes were also awarded for proficiency in Botanical Science, two being offered to the Medical Students in general, and one to the Apprentices of the Society, Mr. J. L. Wheeler being the adjudicator of the awards.

1834

Thus matters went on until the close of the year 1834, when a brother of Mr. Thomas

Wheeler died, leaving a large fortune to be divided among the six sons of the latter. The property was so considerable as to supersede the necessity of any further exertion on the part of these gentlemen to gain their living, and most of them relinquished the public appointments they had previously held, and retired into a position of ease and competence. Among these was Mr. James Lowe Wheeler, who resigned his Lectureship of Chemistry and Materia Medica in a Medical School, and also the Professorship of Botany at the Chelsea Garden. From that period he devoted his leisure time to the indulgence of literary and scientific tastes, to the healthy amusement of yachting, and to the patronage of music, of which he was very fond; but he passed through the usual honourable offices of the Society of Apothecaries, and was also appointed Superintendent of their Chemical Department, in succession to Professor Brande. He became in his turn Master of the Society, and died at a ripe old age in 1870. He left behind him the reputation of being a kind and amiable man, an accomplished scholar, a Naturalist of no mean pretensions, and an excellent Chemist, having learned and taught the science of Chemistry when it was still in its infancy, and being devoted to it to the latest period of his life.

On the resignation of Mr. J. L. Wheeler, it was resolved at a meeting of the Court of Assistants that the best thanks of the Court be given to that gentleman for his valuable services as Professor of Botany during a period of thirteen years, and this vote was accompanied by the sincere wishes of the Court for his long-continued health and happiness.

At the same Court it was resolved that a Committee should be appointed to consider the duties and emoluments attached to the Professorship of Botany.

This Committee was accordingly appointed, and a Report was presented on the 23rd of December, 1834. The recommendations made were founded on those of former Reports, but with some modifications, the chief of which was that in consequence of the great alteration which had taken place in the neighbourhood of the Metropolis within the previous half-century, the same benefit did not result, as it formerly did, from Botanical Excursions, and that the Herborizings be therefore for the future discontinued. It was also suggested that in consequence of the impoverished and unproductive state of the soil of the Chelsea Garden, some means should be taken, and some expense allowed, for improvements in this respect. Some other suggestions were made as to the admission of Teachers of Botany and their Classes into the Garden, but as these suggestions were not adopted, it is unnecessary to refer to them more particularly

---

1835 On the 17th March, 1835, Mr. Gilbert Thomas Burnett was elected Professor of Botany in the place of Mr. J. L. Wheeler, and he commenced his duties in the ensuing summer.

Mr. Gilbert Burnett, F.L S.

Mr. Burnett's career as Professor of Botany, like his life, was a brief and a painful one. He was the last descendant of the celebrated Bishop Burnet, was the son of a medical man practising in London, and was also an Apprentice of the Apothecaries' Hall. He was distinguished in his student career for his zeal in acquiring Botanical knowledge, and by his assiduous attention at the Herborizing excursions he attracted the notice and secured the friendship of Mr. Thomas Wheeler. He became so distinguished for his acquirements in Botany that he was appointed the first Professor of that science in King's College, London, which commenced its career soon after the opening of the University of London (now University College). Mr. Burnett's father dying at an early age, the new Professor was obliged, in addition to his laborious scientific duties, to carry on a general practice for his own support and that of his widowed mother and his sisters. As if these burdens and sorrows were not sufficient for him, his own serious illness supervened, and when in 1835 he was appointed to the Professorship of Botany at the Chelsea Garden, a post for which he was eminently qualified and in which he evidently took great delight, he was already doomed to a premature death. He lectured regularly at the appointed times, and charmed his audiences by his eloquence and his learning, but the labour of public speaking was obviously an onerous one and too much for his strength. Although the Herborizings for the Apprentices had been abolished, he conducted, though on a limited scale, the "Grand Herborizing," riding on a pony, as he was unable to walk, and a few days after the delivery of his last lecture in the Garden in September he was no more. His emaciated figure and frequent cough had indicated for a long time only too plainly the malady under which he was suffering

Mr. Burnett was the author or editor of many works connected with Botany. The principal of these was entitled "Outlines of Botany, including a general History of the Vegetable Kingdom, in which Plants are arranged according to the System of Natural Affinities," in two volumes. This is a most able and laborious work, containing a vast amount of information, both literary and scientific, but rather too abstruse for the use of Students.

Mr. David Don, F.L.S.

In this year the duty of conducting the Examination for Prizes in Botany, previously performed by Mr. James Lowe Wheeler, was confided to Mr. David Don, a distinguished Botanist, for many years Librarian of the Linnæan Society, author of a great number of papers on botanical subjects, and, on Mr. Burnett's death, Professor of Botany at King's College, London. He held the examinership on Botany at Apothecaries' Hall only one year He also died prematurely, being cut off by a malignant disease of his tongue in 1841, aged 41. In the same year, the distinguished Swiss Botanist Decandolle died, aged 63.

At the close of the year 1835, the Court of Assistants met to elect a Professor of Botany in the room of Mr. Burnett, and out of a considerable number of candidates the lot fell upon Dr. Lindley, then in the zenith of his reputation as one of the greatest Botanists in England, or indeed in Europe. He held the office for seventeen years, and exhibited the same energy in performing the duties as he showed in his other public appointments. The present seems to be a fitting opportunity of offering a brief Memoir of this distinguished man.

Dr. Lindley, F.R.S.

John Lindley was born in the year 1799, near Norwich, his father being a nurseryman in good business. He appears to have received only an ordinary school education, and whatever literary acquirements he possessed, and they were considerable, must have been subsequently self-taught. When he left school he at once followed his father's occupation, but he was very early thrown upon his own resources owing to his father's failure in business, and he succeeded to nothing but some paternal debts which he very honourably discharged. In a pecuniary sense, therefore, Lindley began his life under a heavy cloud, and notwithstanding his enormous industry and his great reputation it is doubtful whether he ever was rich.

At the age of twenty he became assistant librarian to the distinguished naturalist and patron of science, Sir Joseph Banks, and at the same early age he published a translation of Richard's "Analyse du Fruit." In 1822 he was appointed Garden Assistant Secretary to the Horticultural Society, and in 1826 he became sole Assistant Secretary, and his connexion with that Society, and their well-known Gardens at Turnham Green, and at South Kensington, lasted throughout his whole working life. In 1829 he was appointed first Professor of Botany in the then University of London, and he held this appointment for thirty years, being associated there with some of the most distinguished teachers of the age; such as Sir Charles Bell, in Physiology and Surgery; Dr. Turner and Professor Graham, in Chemistry; Dr. Grant, in Comparative Anatomy; Dr. Elliotson, in Medicine; Mr. Liston and Mr. Samuel Cooper, in Surgery; and many others. He was a remarkably exact, clear, and impressive lecturer, possessed an admirable faculty of lucid exposition, and was most copious in illustration. It cannot be said that he possessed exactly the gift of eloquence, or that he ever showed any enthusiasm, but as a teacher of Botany in its purely scientific aspects he was unrivalled. Probably owing to the difficulties he had himself experienced in acquiring knowledge, he was minutely particular in his explanations of every fact which could impress the minds of students, and besides the abundant diagrams by which his lectures were illustrated, he took care to draw upon the board any peculiarity of structure he wished to explain, and he wrote out in large letters any unusual word which he employed. In the description of the Natural Orders he was remarkably successful, being abundantly supplied from the Horticultural Gardens with fresh specimens which were distributed to each student and carefully explained. Soon after his appointment as Professor at University College, he published his "Introduction to the Natural System of Botany," a work consisting of 374 pages, dividing all plants into 272 Natural Orders, and which has been already referred to as containing a dedication to the Court of Examiners of the Society of Apothecaries. This work was subsequently expanded into his "Vegetable Kingdom," a large and very able production containing a multitude of illustrations and distributing the Vegetable World into 303 Natural Orders, the third edition appearing in 1853. He published, however, a host of other books, besides monographs and anonymous articles, on almost every branch of botanical science. The list of his published works occupies twelve pages of the large folio catalogue of the Library of the British Museum. Although not the originator of what is known as the Natural System of Botany he was one of its most prominent advocates, and he was undoubtedly the chief means of introducing that system into favour in Great Britain. In his zeal, however, for advancing the new doctrines, he unduly depreciated the merits of Linnæus, and in too many instances wrote and spoke almost with contempt of that great and illustrious naturalist. His own Natural System, although a work of immense labour, shows the difficulty of devising any perfect scheme of botanical arrangement, for he himself altered it several times, and was at last so little satisfied with it as a system that he nearly abandoned in his lectures any consecutive classification of the Orders, and arranged them in arbitrary groups, admitting, as he himself writes in his "Flora Medica," that "no two writers on classification are agreed respecting the exact sequence in which the Natural Orders of plants should follow each other."

His Natural System, which was fully developed in 1853 in his "Vegetable Kingdom," and which perhaps would have been still further expanded if his life had been prolonged, is founded essentially on the Systems of Jussieu and Decandolle, but with considerable modifications, for, whereas both those Botanists make three primary divisions of plants, Lindley increases them to seven, namely, (1) THALLOGENS, or cryptogamic plants without a distinct woody stem, and (2) ACROGENS, or Cryptogams which have a woody stem; (3) RHIZOGENS, an intermediate tribe between Thallogens and Exogens; (4) EXOGENS; (5) DICTYOGENS, plants with netted leaves intermediate between Exogens and Endogens; (6) GYMNOGENS, plants with naked seeds, including the *Coniferæ* and *Cycadaceæ;* and (7) ENDOGENS. These primary divisions are distributed into the 303 Natural Orders comprised in Dr. Lindley's System.

Among other works published by Dr. Lindley was the "Synopsis of the British Flora," published in 1829, being the first attempt to arrange British plants according to the Natural System, and intended to supersede the well-known works of William Hudson, Dr. Withering, Sir James Edward Smith, William Curtis, and Dr. Hooker, which were all arranged on the Linnæan System. (The seventh edition of the last-named excellent work, however, published in 1855 conjointly by Sir William Hooker and Professor Arnott, was arranged for the first time on the Natural System.) Dr. Lindley also published in 1838 his "Flora Medica, a Botanical Account of all the more Important Plants used in Medicine in different parts of the World," but this work, although very laboriously compiled, is of little practical use to the Profession for whom it is designed. Another work was a "Fossil Flora," which he wrote in conjunction with the well-known Geologist, Mr. Hutton, containing an admirable description of fossil plants, with numerous illustrations; and there was one, which it is said he considered his masterpiece, as it contains the greatest amount of original matter, namely, the "Theory and Practice of Horticulture." With the beautiful and fantastic family of the *Orchidaceæ* his name will be always memorably associated, as he described a great number of new genera and species of this curious Order, and wrote a series of works relating to them.

Until he was past fifty he was in the habit of saying that he never knew what it was to feel tired either in body or mind. His first illness was in 1851, and was caused by his arduous duties as a Juror of the Great Exhibition held in that year, but he soon recovered. He undertook, however, against the wishes of his family, the charge of the Colonial Department of the Exhibition of 1862, and, although ailing, he continued at his post to the last. But his mental and physical powers then gave way, and he was compelled to relinquish all active employment, though his bodily health remained pretty good until the 1st of November, 1867, when he died of apoplexy.

In the same year another distinguished Botanist departed this life, but at the advanced age of eighty-one, namely, Sir William Jackson Hooker, author of a great number of botanical works, formerly Professor of Botany in the University of Glasgow, but at the time of his death Superintendent of the magnificent Garden at Kew, now under the management of his no less distinguished son.

In this year, then, Dr. Lindley commenced his duties at Chelsea as Præfectus Horti and Professor of Botany, and as Mr. Don had been elected Professor of Botany at King's College, and was consequently disqualified for holding the post of Examiner in Prizes for Botany at the Society of Apothecaries (see p. 179), the latter appointment was conferred on Mr. Nathaniel Bagshaw Ward, a distinguished Botanist, and a pupil and friend of Mr. Thomas Wheeler. This gentleman held the appointment until 1854. 1836

Dr. Lindley entered upon his new sphere of action with his characteristic energy, both in the delivery of the Lectures, and in the management of the Garden. The Lectures, instead of being delivered once a week during five summer months at ten o'clock in the morning, were now given twice a week during May, June, and July at half-past eight. The arrangement of the plants in the Garden was also thoroughly revised, and Dr. Lindley made many lengthy reports both as to existing defects and as to the best means of improvement. On the 31st of May he reported to the Garden Committee that, having examined the collections of plants in the Garden, he had found them in a very defective state, owing to the want of catalogues, and that he thought this defect should be remedied. Being requested to memorialise the Court of Assistants on the subject, he did so in the following letter:—

To the Master, Wardens, and Court of Assistants of the Apothecaries' Society.

21, Regent St., June 2nd, 1836.

GENTLEMEN,—

At the Meeting of the Garden Committee on Tuesday last, I found it my duty to bring before them the state of the living plants in the establishment, and especially to point out the circumstance that although the collections are extensive and valuable, their utility is almost destroyed by the general want of any catalogue to which Students can refer.

Of the Trees and Shrubs there is no Catalogue.

Of the Glumaceous and Umbelliferous Plants there is no Catalogue, but in some cases written names are attached to the Specimens, without, however, being in all cases correct.

Of the large collection of Herbaceous Plants there does exist a MS. Catalogue, but from a variety of circumstances it has but little real relation to the Plants of which it purports to be a list.

Of the Medicinal Plants there is a correct Catalogue, but these form so very small a portion of what the Garden contains as to affect in no material degree the statement I have made as to the general absence of any means available for Students of profiting by the materials for instruction that the Garden contains.

I suggested to the Committee that under these circumstances it was absolutely necessary for the credit of the Society that means should be taken to ascertain what the Garden contains, and to mark the plants in such a manner that any person visiting the establishment should have the power of ascertaining the names of the species given them. I proposed to superintend the carrying the preparation of a Catalogue into effect; but I stated that it was impossible for me to take upon myself the office of anything beyond superintendence and direction; that the Gardener's time was fully occupied by the avocations connected with his appointment, and that it would be necessary to furnish me with an Assistant till a Catalogue could be completed. I mentioned the sum of 25*l.* as being sufficient to commence the operation with, and probably to complete the Catalogue of the plants not cultivated in the Hot-houses and Greenhouses.

The Committee did me the honour to adopt my views and to recommend them to the consideration of the Court of Assistants, with whom the decision must rest as to whether my plan should be carried into execution.

I therefore have taken the liberty of troubling you with this communication, in the hope that this matter, which I consider a very important one, will meet with your concurrence. Should it be your pleasure to authorize the outlay of the sum of 25*l.*, I hope by the Meeting of the Garden Committee next September, to be able to lay the Catalogue on their table, and then to develope my ulterior plans with reference to bringing the Garden into a more efficient state by the commencement of the next Medical Summer Session.

I have the honour to be, Gentlemen,

Your obedient Servant,

JOHN LINDLEY.

This letter having been read at a Court of Assistants on the 21st of June, 1836, it was resolved, "That the Court cordially accede to the Professor's suggestions in reference to the preparation of a Catalogue, and do order that the required sum be advanced for carrying his recommendation into effect."

The Master stated to the Court that arrangements had been made for the Annual Exhibition of Plants at the Society's General Herborizing taking place at the Hall instead of at a Tavern as heretofore, and that it had been decided to invite to the delivery of the Professor's Lecture on that occasion the Officers of the several Societies connected with Botany, and other individuals eminent in that branch of Science.

1837 In January, 1837, a long Report was read from Dr. Lindley on the condition of the Garden. It stated that the Garden had suffered considerably from the unfavourable atmosphere surrounding it, from the exhaustion and deterioration of the soil, and from the sandy nature of the soil itself. Nevertheless, the Report goes on to state, the Garden still contained a large number of plants well suited for the purpose of instruction, and the number might be increased without much expense. The collections, however, were in great confusion, having been neglected for many years, and in their present state they were almost useless to the Students frequenting the Garden. Dr. Lindley then specifies the particular instances in which the collections are deficient or redundant, an excessive number of duplicates being kept of some, and others being absent altogether which were essential for instruction. He states, moreover, that the square occupied by the officinal plants admitted into the London Pharmacopœia is kept in proper order, but that the plants are not sufficiently numerous. In this communication Dr. Lindley complains of the want of co-operation with his views and wishes evinced on the part of the Gardener. On reading this communication, it was resolved by the Garden Committee, that immediate steps should be taken to carry Dr. Lindley's suggestions and proposals into effect; that a new arrangement of the plants should be made; that duplicates and useless specimens should be removed; that copies of a cheap list of the numbered plants should be published; and that the Gardener should be directed in all respects to carry out immediately the directions of the Præfectus.

In this year it was ordered that the Botanical Season should conclude at the end of July, and that in future May, June, and July should be considered the Botanical Season.

In this year (Aug. 29), it was resolved that it should be recommended that the Botanical Examinations (two in number) should be combined into one, open to all Students, whether Apprentices to the Society or otherwise, and that the prizes given should be three, namely, (1) a gold medal, (2) a silver medal with books; and (3) books only. These recommendations were adopted.

1838 In this year another long and very able Report was presented by Dr. Lindley, complaining that his previous recommendations had not been properly carried out, and commenting very severely on the want of co-operation on the part of the Gardener. The following verbatim extract from this Report admirably expresses what a Botanical Garden should be, and what the Apothecaries' Society desired to make it:—

"The Society of Apothecaries have, with a rare liberality, applied annually a very considerable sum to the maintenance of this Garden for the use of Medical and other Students, and for the purposes connected with Medical Botany. Their object, I conceive, in doing so, has been to form a Garden of instruction which might supply the unavoidable deficiencies of the lecture-rooms of the metropolis, and to afford Students the means of acquiring a practical knowledge of plants in general, and more particularly of species useful in Medicine. Such objects are not effected by little ill-cultivated specimens starved in flower-pots, by crowds of spurious species which have sprung up in German and other Botanical Gardens, and which have no other effect than that of perplexing Students, and throwing Science itself into confusion; or by nameless specimens, flowerless, and almost leafless, illustrating nothing except the endless variety of the Vegetable Kingdom, which requires no illustration. Important purposes such as those I have adverted to are to be accomplished in a very different way. They require well-cultivated specimens, illustrating remarkable peculiarities of organization, anomalous forms of vegetation, representatives of Natural Orders, and especially species known to possess energetic properties either poisonous or curative. For such a collection the Society's Garden, with the existing houses, is sufficiently well adapted; the locality offers no serious impediment to its preservation, and instead of requiring an additional annual outlay, I have no doubt of its being well maintained at an expense considerably less than that which is at present incurred."

The Report goes on to animadvert very strongly upon the conduct of the Gardener in opposing the writer's views, and it concludes by asking the Committee to declare distinctly whether they adopt his (Dr. L.'s) views, or those of the Gardener. The Committee thereupon resolved that Dr. Lindley's views should be adopted, and that the Gardener be directed to follow out in all respects the directions and orders of the Præfectus.

On the 28th August of this year, Dr. Lindley reported that the resolutions of the last Committee were being carried into effect, but that little progress had yet been made in marking the plants with their names. It was ordered that a new arrangement of the plants in the Garden should be made under the direction of Dr Lindley, and that the Medical plants should be all marked before the next spring.

1839 On the 28th of May in this year it was reported by the Præfectus that the new arrangement of the Medical plants had been completed as far as possible, and that the Society had received great assistance from the Messrs. Loddiges of Hackney, in giving several plants.

1846 On the 27th May in this year a limb from the cedars of Lebanon was blown off, and it was ordered by the Garden Committee that it should be converted into four chairs for the use of the authorities of the Hall, and this was accordingly done. A chair had been constructed of a branch previously blown off in 1812, and it bears a suitable Latin inscription.

On the 16th of October of this year it was announced that Mr. William Anderson, the Curator of the Garden, had died since the last meeting at the age of eighty, and that he had been buried between the graves of Sir Hans Sloane and Mr. Philip Miller.

Mr. William Anderson, F.L.S.

William Anderson was a man of humble extraction, and was born in Scotland. He was brought up as a gardener, and officiated in that capacity all his life, but he had a considerable knowledge of Botany, particularly of indigenous plants, and was attached to the doctrines of Linnæus, and either from advancing years, or early associations, was not much inclined to favour the promotion of the new Natural Systems. He held the office of Gardener (this title being changed during his lifetime to that of Curator) of the Chelsea Garden for thirty-two years, and although he was a man of rather rough manners, he was of a generous disposition, and did many kind acts for necessitous friends. He lived alone, and had no known relatives, those he had being discovered only after his death, in consequence of an advertisement inserted in the papers announcing that he had died intestate, and calling upon any of his kindred to come forward and claim what little property he had left behind him. In consequence of this announcement some distant relations presented themselves from a remote part of Scotland and obtained possession of his effects.

A singular instance of his kindness, and of the unostentatious and self-denying way in which he sometimes exercised it, is the following. He possessed a diamond ring, of the value of one hundred guineas, which had been presented to him by the Emperor of Russia, in return for the care he had bestowed on some valuable orange-trees belonging to His Majesty. Those who remember Mr. Anderson's tall and burly form and his ordinary coarse and old-fashioned style of dress, will be inclined to wonder, first, how a diamond ring of such value would be consistent with the rest of his attire, and secondly, what kind of ring it must have been to fit the gigantic finger of the wearer. The ring, however, was constructed specially for the purpose intended, and Mr. Anderson was of course very careful of it and proud of possessing it, although he very seldom wore it. But, on his death, this valuable ornament could nowhere be found, and after ransacking all his drawers and cabinets in vain, a pawnbroker's ticket was at last discovered showing where the costly trinket had been pledged. It turned out on investigation, that Mr. Anderson had been induced on some occasion to help a necessitous friend, and not having the money sufficient to enable him to do so, had actually pawned this keepsake for the purpose! The ring was forthwith redeemed, and the difference in value was added to the other few effects left by Mr. Anderson, and the whole produce was divided, as before stated, among his distant relations. It does not appear that the friend, whoever he was, for whose benefit the ring was pledged, ever repaid the loan.

Mr. Anderson was a Fellow of the Linnæan Society, and he wrote several papers on horticultural subjects in the periodicals devoted to gardening.

1846

In order to supply the place of Mr. Anderson the Garden Committee asked the advice of Dr. Lindley as to the choice of a successor, and Mr. Robert Fortune was recommended and eventually elected to the office of Curator of the Garden. The choice was a particularly happy one, for Mr. Fortune had held an important position as Botanical Collector to the Horticultural Society of London, and had just returned home after a three years' exploration of the Botany of China, from which country he had brought back a great number of valuable specimens. The record of his travels has been given to the world in a most interesting work entitled "Three Years' Wanderings in the Northern Provinces of China, including a Visit to the Tea, Silk, and Cotton Countries, with an account of the Agriculture and Horticulture of the Chinese, New Plants, &c."

This book not only displays extensive Botanical and Horticultural knowledge, but it also contains a great amount of interesting anecdote, and amusing descriptions of the manners, customs, and social life of the Chinese people, together with some excellent engraved illustrations of striking scenery.

Mr. Fortune's name has been perpetuated in the name of a tree of the walnut tribe (*Juglandaceæ*) brought by him from China, the *Fortunæa Chinensis*, and in the specific names of many other plants, among others the *Chamærops Fortuni*, brought by him from the island of Chusan, and a fine specimen of which is now to be seen in one of the green-houses of the Chelsea Garden.

1847 Mr. Fortune, under Dr. Lindley's superintendence, exerted himself most energetically in re-arranging the plants in the Chelsea Garden, and in introducing such other reforms as the advanced state of Botanical science demanded. In carrying out the changes thus deemed necessary, fresh expenses were unavoidably incurred, and a new subscription was called for at the hands of the members of the Apothecaries' Society. The actual outlay on new buildings and other improvements amounted to 1220*l.*, and a considerable sum was raised by private subscription among the members to discharge the debt thus incurred, but a considerable deficiency remained which was supplied by the Corporation of the Society, who voted 700*l.* for the purpose.

During Mr. Fortune's Curatorship two new span-roofed glass structures, a stove, and a greenhouse, were built and heated by the then new Polmaise, or hot-air system. This plan, however, did not succeed, and it was accordingly abandoned a year or two afterwards, and a hot-water apparatus was substituted.

1848

Mr. Fortune's services as Curator of the Garden were of short duration, for he soon severed his connexion with the Apothecaries' Society under the following circumstances, which were very honourable to himself and to Horticultural science in general. On the 8th of May, 1848, Dr. Forbes Royle, on the part of the Court of Directors of the East India Company, addressed a letter to the Garden Committee, requesting them to allow Mr. Fortune to proceed to China for the purpose of collecting the best kinds of Tea-plants and seeds in the northern parts of that Empire, with a view of cultivating the Tea-plant in the Himalayan mountains. Dr. Royle requested that facilities might be given to Mr. Fortune for carrying out this object, and the latter gentleman addressed a communication to the Garden Committee expressing his willingness to accept the offer made to him by the East India Company, and asking for two years' leave of absence from his duties at the Chelsea Garden.

The Committee, after due consideration, came to the resolution that, regarding the interests of the Garden, they were unable to grant the required leave of absence to Mr. Fortune, but that they placed no obstacle in the way of that gentleman accepting the offer made to him. Thereupon Mr. Fortune, in very courteous terms, tendered his resignation, which was accepted, with an expression of regret at the loss of his services, which had at all times given the greatest satisfaction, and with the best wishes for his future success and prosperity.

It may not be irrelevant to mention that

Mr. Fortune's mission was attended with great success both as regards himself and the objects for which his travels were undertaken, and that he is now living in comfort on the well-earned proceeds of his talents, industry, and enterprise.

On the 29th of May of this year Dr. Lindley was requested to name a successor to Mr. Fortune as Curator of the Garden, and he nominated Mr. Thomas Moore, who was subsequently elected, and who has continued to hold that office from that time to the present.

Here it may be interesting to record the annual expense of the Garden for twenty-seven years.

		£	s.	d.
1850,	the expenditure was	£856	6	0
1851,	„	574	18	1
1852,	„	671	10	1
1853,	„	522	2	7
1854,	„	648	10	5
1855,	„	443	5	10
1856,	„	320	4	5
1857,	„	332	7	5
1858,	„	457	10	8
1859,	„	313	4	11
1860,	„	272	19	3
1861,	„	273	14	11
1862,	„	677	17	3
1863,	„	766	14	1
1864,	„	447	4	9
1865,	„	721	14	1
1866,	„	369	16	7
1867,	„	377	14	10

1868,	„	£698	16	8
1869,	„	464	13	10
1870,	„	463	9	10
1871,	„	415	13	0
1872,	„	460	2	7
1873,	„	702	1	8
1874,	„	683	6	11
1875,	„	604	14	11
1876,	„	433	5	5

1853 In this year some very important steps, but of a retrograde character in reference to the science of Botany, were taken in the affairs of the Chelsea Garden. It will be observed, on referring to the early part of these Memoirs, that the freehold property on which the Garden is situated belongs to the noble family of Cadogan, and that by the will of Sir Hans Sloane it is allowed to be used as a "Physick Garden," under the management and at the expense of the Society of Apothecaries. In case of this body declining or neglecting to maintain it, or applying it to any other than scientific purposes, it will be seen that the Royal Society and the Royal College of Physicians of London may successively become the occupiers of the ground on the same conditions as those imposed upon the Society of Apothecaries, but if both these bodies decline the offer, then the property reverts to the descendants of Sir Hans Sloane, who are now represented, as far as the Garden is concerned, by the Earl of Cadogan. It is perhaps unnecessary to mention, after what has been already written in these pages, that neither the Royal Society nor the Royal College of Physicians (except a donation from the latter of 100*l.* in the year 1723) have ever contributed anything to the support of the Garden, and that neither of those bodies has ever shown any desire to become possessed of it; and it may also be mentioned that the Cadogan family have never manifested any interest in the maintenance of the Garden as a scientific institution. Hence the whole expense of its support has fallen exclusively upon the Society of Apothecaries, who, except in reputation, have never received anything in return.

These preliminary observations having been introduced, it is only necessary to state that in the present year (1853) some correspondence was instituted with Lord Cadogan and others having reference to the probable relinquishment of the ground as a Botanic Garden, and during the negotiations it was thought expedient that the expenditure of the Garden should be reduced to the lowest possible point.

A series of sweeping changes were therefore proposed, the minute respecting them bearing date 27th June, 1853, and most of them were carried into effect. The chief of these proposals were that the Lectures at the Garden should be discontinued, that the office of Præfectus Horti should be abolished, and that no plants should be cultivated at the Garden except such as could grow without artificial heat, the object of course being to save the expense of fuel. Other economical reforms were suggested, such as the discontinuance of the dinners which had previously been held in the Garden, the abolition of the Fees for attendance on Garden Committees, the dismissal of labourers, and the suspension of the annual award of Prizes in Botany; and it was now proposed that only a fixed sum of 250*l.* per annum, to be paid out of the Corporate funds of the Society, should be devoted to the future maintenance of the Garden.

Thus the services of Dr. Lindley were dispensed with, the summer Lectures, which had been regularly delivered for twenty-three years, were discontinued, and the establishment was left under the sole care of Mr. Moore, the Curator. Some of the plants were sold, some of the more delicate ones were exchanged for hardy species, many of the hothouse and greenhouse plants which had been collected during many previous years were given away, and one of the new glass-houses was taken down and sold. But the whole of the recommendations were not adopted, for the annual Prizes were still awarded, and, although the dinners at the Garden were abolished, the Garden Committees after a short interval resumed their proceedings.

If any prospect, however, had been entertained of effecting some great immediate economy in reference to the maintenance of the Garden, the idea was rudely dispelled by an accidental occurrence which took place in the autumn of this year. On the 18th of October, 1853, at a Meeting of the Court of Assistants, it was announced that a sewer was about to be constructed along the road which runs by the north side of the Garden, where the large lecture-room and the Curator's residence were situated, and that this construction would render the whole building in the Garden unsafe by undermining the foundations. It was therefore determined that the old building, which was in a decayed state from lapse of time, should be pulled down, and that on the site of the old lecture-room a new residence should be built for the Curator. These alterations were accordingly made, and the new building, as it now exists, was erected, but the expenses were considerable, amounting to upwards of 300*l.*; and further expenses were subsequently incurred by the building of a committee-room, with offices adjoining, and the whole outlay, which was spread over several years, amounted to between 700*l.* and 800*l.*

In the autumn of this year Mr. Nathaniel Bagshaw Ward was elected Master of the Apothecaries' Society, and he made his year of office memorable by his efforts to develope the scientific features of the Society, and especially by giving two large *soirées*, to which all the scientific world was invited, and in which the exhibition of microscopical objects formed the principal attraction, although Mr. James Lowe Wheeler also contributed a beautiful series of chemical and electrical phenomena and experiments. Mr. Ward now relinquished the post of Examiner for Prizes in Botany, which he had held for seven teen years, and he was succeeded in the duty by a gentleman who has since attained the highest eminence in science, namely, Dr. Hooker, now Sir James Dalton Hooker, lately President of the British Association for the Advancement of Science, and at present (1877) President of the Royal Society, and since the death of his father, Sir William Hooker, Superintendent of the Kew Garden.

For about nine years after 1853 very little was done for the improvement of the Garden as a place of study, and a variety of proposals and suggestions in reference to the future destination of the property were considered or entertained, and a number of inroads on the part of Railway Companies were threatened; and these circumstances altogether militated against any attempt at the effectual promotion of Botanical Science, although the Garden was still visited by the Medical Students of the Metropolis for the purpose of instruction. But after this interregnum of what may be called the dark age of Botany at the Chelsea Garden, and when it had been found that the various proposals as to the alienation of or encroachment upon the property had fallen to the ground, the Executive of the Apothecaries' Society reconsidered the whole question as to the use and destination of the establishment.

At a Court of Assistants, held on the 31st of July, 1862, a communication was received from the Garden Committee, who suggested the necessity of improving the Garden, of rendering it better adapted to the purposes for which it was intended, and of cultivating it more efficiently; and it was thereupon resolved that a sum of 700*l.* should be placed at the disposal of the

Garden Committee for the purpose of carrying out the recommendations with regard to the repairs and improvements now rendered necessary.

On the 31st of October of this year, at a Meeting of the Court of Assistants, it was resolved that the Herbarium existing at Chelsea Garden should be presented to the Trustees of the British Museum, and this was accordingly done.

The Herbarium, or rather Herbaria, for there are three, thus handed over to the British Museum, forms a very valuable series. The chief part of it consists of the original herbarium of the distinguished British naturalist, John Ray, who bequeathed it to his friend Samuel Dale, already mentioned in these Memoirs (p. 66), who, at his death in 1739, left it, together with his own books and plants, to the Apothecaries' Society. It was preserved in suitable presses in the Chelsea Garden, under the direction of Sir Hans Sloane, as previously described in the above-mentioned page. Isaac Rand, a former Demonstrator of Botany at Chelsea, also made an extensive *hortus siccus*, which, at his death, was placed together with those of Ray and Dale, and these are the three herbaria now referred to. The herbarium of Ray, which is in good preservation, is contained in nineteen thin quarto fascicles, the plants being sewn on the paper, and labelled in the peculiarly neat and plain handwriting of that illustrious botanist, and the specimens are accompanied by a manuscript index, entitled "Horti Sicci Raiani Catalogus." *The Journal of Botany* for the year 1863, writing of this collection, then lately deposited in the British Museum, says that "the importance of this collection in determining precisely what are Ray's species cannot be over-estimated," and it goes on to state that together with the herbaria of Dale and Rand, and those of Sloane, Sherard, Petiver, and others, it will materially assist the preparation of a Report, then in contemplation, on "The Plants of Ray's Synopsis Stirpium." In a note on the subject of Ray's Hortus Siccus in the same Journal for 1870, by Mr. Henry Trimen, M.B., F.L.S., the writer adds his testimony as to the great value of the collection, and points out the localities from which most of the specimens were probably obtained, Ray having made collections of plants both in foreign countries and in Great Britain. He gives the modern names of several British plants thus preserved and described by Ray, and it is very interesting to compare the simplicity of the Linnæan or modern nomenclature with the accurate but cumbrous terms used by Ray, who, it may be mentioned, died two years before Linnæus was born. Thus, to quote only a single instance, the "Juncus parvus montanus cum parvis capitulis luteis" of Ray, becomes, in the short and simple Linnæan nomenclature, "Scirpus cæspitosus."

It was a happy thought to transfer these valuable herbaria from the sheds in Chelsea Garden, where they were gradually decaying under the influence of time, weather, and neglect, to the spacious rooms of the British Museum, where they are safely protected, and where they may be seen and appreciated by all lovers of Natural History. It should be mentioned that, under the care of the present superintendents of the Botanical Department of the Museum, Mr. Carruthers and Mr. Trimen (1878), the specimens have been preserved from further decay, and are mounted on fresh paper. This transfer of the herbaria was in great measure due to the influence and the exertions of Mr. Ward, to whom botanical science is so much indebted in this as in many other respects.

1863 The Court of Assistants now determined to take active steps in improving the Garden, and in the words of a printed appeal by the Master and Wardens to the Members of the Society, published in January, 1863, "when they reflected how much benefit the Garden had conferred in times gone by, with what pride it had been cherished by their predecessors, and, above all, when they found how numerous a body of Medical Students were still anxious to profit by it (no fewer than 500 Students having applied for admission during the past summer) they unanimously resolved that the Garden should be continued, and that a vigorous effort should be made to render it as efficient as possible for the fulfilment of the important objects to which it has been so long devoted."

In order to carry this resolution into effect it was determined to make several new arrangements at the Garden, as for instance, to make a new and complete collection of all medicinal and economic plants; to enlarge the collection of the more hardy herbaceous plants classified according to the Natural System; to construct some new buildings and to rearrange others; to introduce examples of Wardian Cases, in illustration of the utility of this contrivance in the cultivation of plants; and to effect other important changes.

Mr. Nathaniel Bagshaw Ward, F.R.S., F.L.S.

Among the members of the Society who took the most prominent part in restoring the Chelsea Garden to its former exalted position as a school of practical Botany was a gentleman whose name has been already mentioned, Mr. Nathaniel Bagshaw Ward, who, although his modest and retiring character prevented his appearance in public either as an author or as a Lecturer, was one of the most accomplished Botanists of his day. No excuse is necessary for the introduction of the following brief Memoir of this accomplished and excellent man, who is well remembered by many surviving friends, and whose memory must be dear to every member of the Apothecaries' Society.

He was the son of a medical practitioner in the east-end of London, and when quite young he exhibited a great love for natural history, and made little collections of plants and insects. It seems that he contracted some boyish inclination for a sea-faring life, and his father indulged his fancy by sending him, when thirteen years of age, on a voyage to Jamaica. While the vessel in which he sailed was lying off the island he was delighted in admiring the sea-birds, the dolphins, and the flying fish, and on the edge of the coral reef which forms the outer margin of the harbour he was still more delighted by the beauty of the animal and vegetable forms growing in the water. When he visited the interior of the island the superb tropical vegetation, especially the Palms and Ferns, made an impression on his mind which was never effaced, and he became from that time an ardent and devoted Botanist. But the voyage home quite cured him of his love for a sailor's life, and on his return to England he was apprenticed to the medical profession, and it may readily be supposed that under the guidance of Mr. Thomas Wheeler, at the Chelsea Garden and in the Herborizing Excursions, his love for Botany was still further developed and matured. His medical education was pursued at the London Hospital, where he zealously performed all the duties imposed upon him in that great establishment, paying special attention to the surgical cases and often acting as dresser. It is almost needless to remark that he passed creditably the necessary medical and surgical examinations, and became in due time a member of the College of Surgeons and of the Apothecaries' Society. He practised for a great number of years in Wellclose Square, in the east of London.

The early lessons inculcated by Mr. Wheeler at the Herborizings inspired him with such a love of Botany that, without in any way interfering with his medical practice, he took every possible opportunity of indulging in this favourite pursuit, and he accomplished his object by stealing from the day some of the early hours usually devoted to slumber, and some of the later ones usually devoted to ease or recreation. He was often out alone by sunrise in the summer time at Shooter's Hill, or Wimbledon, or elsewhere in the rural districts around London, where he collected all the plants he could find, and was back home and ready at the usual time for the professional business of the day; or, when business was over, he would sally forth to Kew Gardens or the Chelsea Garden, or to the Messrs. Loddiges' well-known Garden at Hackney, and so finish the evening. When a family had gathered round him, and they were all staying, as they frequently were, at some favourite spot in Kent, he and his sons would often be up before sunrise among the brake, and the heather, and the wild flowers, and after two or three hours' search for plants they would all breakfast at some wayside inn, and after some more plant-hunting would return when the day was far advanced. Cobham Park, with its magnificent scenery of forest-trees, was a favourite resort, and it was Mr. Ward's great delight to come here at sunset, when the sky overhead was almost darkened by the thousands of birds,—rooks, jackdaws, and herons,—coming home to their resting-places and making the air resound with their exuberant though far from melodious cries.

The residence of Mr. Ward in the closely peopled neighbourhood around Wellclose Square was the cause of the happy discovery he made as to the preservation of plants in closed cases. There was a little garden in front of his house, with a vine running up the wall, and there was every possible contrivance about the windows for the reception of plants; there were also saxifrages on the roof, fig-trees and Virginian creepers on the back of the residence, and a small but well stocked garden in the rear. But this appearance of vegetable life and vigour was maintained with the utmost difficulty in the smoky atmosphere of the locality, the soot and dust clogging up the lungs of the plants and impeding their respiration, the cold dry winds carrying off the vapour from the leaves and the moisture from the subjacent mould, and the deleterious gases around poisoning all the tissues. The only resource Mr. Ward possessed to preserve the semblance of verdure was to repair to the country or to nursery-grounds, and bring back a fresh supply of plants and flowers. But his difficulties were eventually removed by a very simple method, which, however, had never struck the minds of his predecessors.

In the summer of 1829, he had placed the chrysalis of a moth in some mould in a glass bottle covered with a lid, in order to obtain a perfect specimen of the insect. After a time a speck or two of vegetation appeared on the surface of the mould, and turned out to be a fern and a grass, which continued to grow in this closed vessel. On reasoning out the matter, he found that in the bottle there existed the necessary conditions of vegetable life, namely, air, light and moisture, while deleterious influences were excluded. This simple experiment was followed by many others, and the Wardian cases (as they are called) are now so well known as to supersede any necessity of description. In 1836, Mr. Ward announced his discovery in a letter to the late Sir William Hooker; in 1838, the late Mr. Faraday lectured on the subject to a large audience at the Royal Institution; in 1842, the first edition appeared of Mr. Ward's work "On the Growth of Plants in Closely-glazed Cases;" and a second edition appeared some years afterwards, with illustrations by the present distinguished artist, Mr. E. W. Cooke, R.A., and his sister the late Mrs. S. H. Ward. A second lecture on the subject was delivered at the Royal Institution in 1854 by Dr. S. H. Ward.

The applications of the "Wardian Cases" are so numerous that only a few of them can be mentioned in these pages. Among those universally known are the facilities afforded by this contrivance for the growth of plants in the dwellings of all classes, and the preservation of verdure and freshness under every kind of deleterious influences in the atmosphere; and for the preservation of some animals the plan is likewise adapted, the *Vivaria*, as they are called, being modifications established by Mr. Ward himself. But the application of the greatest importance to horticulturists is for the transport of plants to and from distant countries. Under the old plan plants were either packed in boxes or allowed to grow during the voyage, exposed to various injurious influences, as sea-water, storms, varying degrees of temperature, &c.; and the consequence was that a large percentage died in the transit. Mr. Fortune, already mentioned in these Memoirs, who brought over a great number of plants from China to this country, states that out of 250 plants put into the cases by him in China, only thirty-five died during the voyage; while in 1819 it was stated in a communication to the Transactions of the Horticultural Society that at that time *only one plant in a thousand* survived the voyage from China to England. Mr. Fortune also conveyed in these cases 20,000 Tea-plants from Shanghai to the Himalaya mountains, and, since that feat, the Cinchona tree, a native of South America, has by the same agency become established in India.

When residing in Wellclose Square, Mr. Ward gave frequent *soirées*, at which the microscope and its revelations were the prominent features. Out of these meetings sprang the Microscopical Society in 1840, Mr. Ward, the late Dr. Bowerbank, and the late Messrs. Jackson and Quekett being its principal founders.

Throughout the greater part of his life Mr. Ward was associated with the Apothecaries' Society. He held the office of Examiner for Prizes in Botany from the year 1836 until 1854, in which year he became Master, when he endeavoured to bring the scientific element of the Society into prominence by giving on a very large scale at the Hall, two microscopical *soirées*, which have never been surpassed either there or elsewhere; and he ultimately filled the responsible post of Treasurer. He was always best known in the Society as the strenuous advocate of scientific progress, and he was eminently conspicuous in the management and improvement of the Chelsea Garden, especially during the few years before his death, when many beneficial changes, about to be referred to, were effected at his suggestion and under his superintendence.

In 1856 a large number of friends combined to recognize Mr. Ward's services by having his portrait painted by J. P. Knight, R.A., and placed in the rooms of the Linnæan Society at Burlington House, and a portrait of him is also placed in one of the large rooms of the Apothecaries' Hall. He never received any recognition of his services from the State, but he was compensated for this neglect by the love and friendship of a large circle of friends, and by reflections such as the following passage from the *Spectator* conveys to the mind, and which is transcribed from his own handwriting in an interleaved copy of his little book. "The consciousness of approving one's-self a benefactor to mankind is the noblest recompense for being so: and undoubtedly the most interested cannot propose anything so much to their advantage."

During the latter years of his life Mr. Ward resided at Clapham Rise, the house where he lived being now occupied by one of his married daughters, Mrs. Braithwaite; and those who visited him there (and his house was open to all) will retain a vivid recollection of his unpretending hospitality, as well as of the many botanical attractions the mansion presented, whether in the closed cases, or in the large Fern-house, or in the open air in the garden.

About the beginning of the year 1868, his health began to fail, and in May he went to St. Leonard's in hope of recovery, but he died there quite suddenly on the 4th of June, aged 77, and was interred at Norwood cemetery in the presence of a large number of friends, among whom were many representatives of the Apothecaries' Society, whom he had served so long and so well.

Mr. Ward, as a medical practitioner, was an honour to his profession, and as a naturalist was a man of superior attainments. He was an excellent classical scholar, and although his modesty seemed to prevent him from making any public displays of oratory he was a ready and fluent speaker. He was a genial companion, full of anecdotes, which he related with great felicity, and he was not only cheerful himself but imparted his cheerfulness to those around him. A Journalist, whose friendship he enjoyed for many years, in noticing his death spoke of him as "one of the gentlest, kindest, and purest of human beings," and to this it may be added, partly from general and partly from personal knowledge of his character, that he was true in all the relations of life, and was a genuine and practical Christian. His life is a striking instance of the gentle influence imparted to the mind of man by the early love and contemplation of nature, a feeling which never forsook him to the last, and which, it may be moreover remarked, sustained him through many and great domestic sorrows, not the least of which was the loss of a son of very great promise, who was cut off in the very spring-time of fame and prosperity.

We cannot close this brief memorial without quoting the following passages from the pen of Dr. (the present Sir Joseph Dalton) Hooker, written on the occasion of Mr. Ward's death. "Mr. Ward, whom I knew for full thirty years of my life, was as warm and steady a friend as I ever possessed; and it would be difficult to say which of the many excellent traits of his estimable character was most worthy of imitation, his love of truth, or his appreciation in others of generous qualities far inferior to his own; his unselfish regard for the happiness of those around him; or the absence of all vanity, littleness, and self-love; or his eager desire to promote the worthy aspirations of the young, and administer to the failing faculties of the old. He sought the acquaintance of youths, especially for the purpose of fostering the tastes of those who took to natural history, and instilling the love of Nature into those who did not, with the one object in both cases of rendering their lives the happier; and he extended the same solicitude to the poorest of the poor with a zeal and singleness of purpose which would have appeared morbid in a man of less cultivated tastes or less scientific acquirements . . . . In the memory of those who knew him, Mr. Ward will live as a type of a genial, upright, and most amiable man, an accomplished practitioner, and an enthusiastic lover of nature in all its aspects."

1863 The Court of Assistants having determined to carry out energetically the proposed alterations and improvements in the Chelsea Garden, the necessary orders were given, and 500*l*. were voted to defray the expenses. A provisional list of medicinal and other plants proposed to be cultivated in the Garden was prepared by the Curator, Mr. Moore, with great care, and was printed for distribution, the object being to induce the Members of the Society and others to assist in the completion of the various collections of plants by contributing specimens, either obtained by themselves or through the agency of correspondents or relatives abroad. This provisional list comprised a great number of species, and was arranged, in the main, on the principles of Lindley's Vegetable Kingdom and the Flora Medica of the same author, but references were also copiously made to Pereira's Materia Medica, Royle's Manual of Materia Medica, the Pharmacopée Universelle, and other sources. Mr. Moore evidently wished to make the Chelsea Garden as complete a repertory of Medical Botany as possible, and such as should be available for instruction by Professors of Botany and their students, who were invited to make use of the establishment in its renovated form. The design of this list was to make suggestions as to the plants which might be contributed by correspondents at home or abroad, and in response to the appeal several contributions were received. Although the results did not altogether meet the expectations or the requirements of the occasion, still some great and important changes were effected, not only in the addition of new plants, but also in the construction of new and improved buildings for the protection and cultivation of the more tender species, and in the furnishing of the rooms of the Assistant Gardeners with suitable books and specimens for the instruction of these officials.

In all these improvements, the services rendered by Mr. Nathaniel B. Ward were eminently conspicuous, that gentleman being at the time now alluded to, and having been many years previously, an active Member of the Garden Committee. The Garden was now placed altogether in a much better condition; hot-house plants were again taken into cultivation; and a new structure, being a kind of Wardian Case on a large scale, was built for the purpose of keeping such plants as, although hardy, do not well bear the climate of London. Under the directions of Mr. Ward, too, the plants in the two lean-to greenhouses on the north side of the Garden were planted out on rocky mounds, and a new and more extensive collection of plants possessing medicinal properties, or more or less employed medicinally, was at the same time formed near the entrance-gate, and the whole Garden underwent a process of renovation, which added most materially both to the beauty of its appearance and to its efficiency as a school of Botany. Indeed, from that time to the present, the establishment may lay claim to the possession of the most complete set of hardy medicinal plants of any of the London gardens, although, from the more ample accommodation and the more abundant means of acquiring rare plants possessed by some other institutions, the indoor collections at Chelsea are not in the same advanced condition.

1866

In this year Dr. Hooker resigned the appointment of Examiner for prizes in Botany to the Apothecaries' Society, and the Rev. Miles Berkeley, M.A., F.L.S., was elected to fill the vacancy, and he has held the post from that time to the present.

The opportunity here seems to be a suitable one for observing that prizes in Botany have been regularly awarded by the Society every year since 1830, and the recipients of these prizes are duly recorded in the annual lists published by the authorities. Among the names of the successful candidates, the following may be adduced as showing that the competition for these rewards was among the early struggles of many who have since arrived at distinction, either in the profession of Medicine, or in general science :—William Jenner, Golding Bird, Philip Bernard Ayres, William Tegetmeier, George Johnson, Thomas Henry Huxley, John Syer Bristowe, Frederick William Headland, Maxwell Tylden Masters, John Harley, Charles Hilton Fagge, Henry Trimen, Henry Greenway Howse, Henry Charlton Bastian,

Henry Power. Some of these are no more, although their names will be honourably remembered; it would be invidious to select any of the living for special notice, and indeed their merits and their position are too well known to require any such specification.

1870 In this year a notice was served on the Executive of the Apothecaries' Society, to the effect that the Metropolitan Board of Works were about to construct an Embankment along the northern bank of the Thames, extending from Pimlico to Battersea Bridge, and that certain encroachments would be made on some part of the Chelsea Garden.

In order to explain the altered conditions which this construction would involve in reference to the Garden, the River, and the adjacent bank, it is perhaps necessary to mention that in former times the Garden communicated directly with the River, a wall running along the river side, and a gate opening upon some steps from which passengers could embark and disembark when visiting the Garden. At high water, the river almost and sometimes entirely washed the Garden wall, and at low water a considerable space intervened between the

Garden and the stream. The general and public objects of the Embankment were manifold, and some of them are very obvious, namely, to contract in some measure the channel of the river; to reclaim a large extent of foreshore for the purpose of a road and in order to erect public or private buildings, and generally to add to the ornamental appearance of the Metropolis; which last object has undoubtedly been achieved by the magnificent Embankments of which London has now to boast.

By carrying out the proposed plan, the occupiers of the Chelsea Garden would lose their immediate access to the river and would sacrifice their right to their portion of the foreshore, and a road would intervene between the Garden and the river. Hence it became necessary to settle by legal arrangement the relative rights and claims of the Apothecaries' Society as tenants of the Chelsea Garden on the one hand, and the requirements of the Metropolitan Board of Works on the other.

The result of the correspondence and of the legal discussions and decisions, was that under the powers contained in the Thames Embankment (Chelsea) Act 1868, the Metropolitan Board of Works acquired, for the purpose of the Chelsea Embankment, the water frontage of the Society. But, in addition to other valuable rights which the Society obtained from the Board of Works by way of compensation, a large sum of money was paid to them by the Board for the erection of a handsome wall, railing, and entrance-gates, facing the Embankment.

The Society have in addition expended a considerable sum from their own funds in constructing an ornamental slope or bank behind the new wall.

1874 On the 9th of May of this year the Chelsea Embankment was opened to the public, the ceremony being inaugurated by his Royal Highness the Duke of Edinburgh and his then newly married bride. The Court of Assistants of the Apothecaries' Society invited a large party to view the procession from the Garden, and as the weather was fine the scene afforded much pleasure to the assembled company.

1876 During this year, while the improvement and enlargement of the Garden were in progress in connexion with the completion of the Chelsea Embankment, it became known to some Members of the Court of Assistants that the Director of a large establishment at Chelsea for the education of young women as teachers had requested the permission of access to the Society's Garden, for the purpose of instruction in Botany. The Court at once acceded to the request, and the result was found to be highly satisfactory, large numbers of these young women, accompanied by their teachers, having since this period visited the Garden, and having derived great benefit from the opportunities for study thus afforded.

1877

It was consequently determined by the Court of Assistants to encourage still further the prosecution of botanical study by young women, and the Court, in June, 1877, resolved that they would offer to candidates of the female sex prizes for proficiency in Botany of the same value and to be competed for in the same manner as in the case of Medical students. In passing this resolution, it was not at all intended to promote the assumption by ladies of medical titles, or to sanction the adoption of the medical profession by females, but merely to encourage in that sex the acquisition of botanical knowledge as a branch of general science. The authorities of Kew Garden, especially Sir Joseph Hooker, who had formerly held the post of Examiner for Botanical prizes at Apothecaries' Hall, and Mr. Thiselton Dyer, the Assistant Superintendent at

Kew, were communicated with, and the project received their warm support.

With a view, therefore, of promoting among young ladies the acquisition of accurate knowledge in a branch of Natural History so entirely consistent as Botany is with female tastes and pursuits, but without any reference whatever to the connexion of this science with Medicine, a Committee was appointed to prepare a scheme for the proposed competition, and the results were embodied in the following published paper:—

EXAMINATION FOR PRIZES IN BOTANY FOR YOUNG WOMEN.

REGULATIONS.

THE SOCIETY OF APOTHECARIES of London offer to young Women Students of Botany, for proficiency in that Science, Prizes to be competed for under the following conditions:—

The Competition will be open to all young Women who shall produce from their teachers certificates that their age at the time of Examination does not exceed 20 years.

The Examination will be in General and not Medical Botany.

The Examination will consist of questions both written and oral, in—

1st, Structural Botany,
2nd, Vegetable Physiology,
3rd, Description of Living Plants,
4th, Systematic Botany,

so far as those subjects are contained in Sir Joseph Hooker's "Science Primer"—"Botany," and in Professor Oliver's "Lessons in Elementary Botany."

The first Examination will take place in London, on the third Wednesday and the third Friday in June, 1878.

A portion of the second day's Examination will be devoted to Microscopical demonstration.

Certificates of Merit will be awarded, in the first instance, to a certain number of Candidates to be determined by the Examiner; and such selected Candidates will be allowed to compete for the Prizes, consisting of the Society's Gold Medal to be awarded to the first in order of merit, and of a Silver Medal or of Books to the second.

Candidates will be required to send their names and their residences, at least fourteen days before the Examination, to the BEADLE, Apothecaries' Hall, Blackfriars, E.C., when they will receive tickets of admission to the Examination.

Copies of the Regulations may also be obtained by application to the BEADLE as above.

By order of the Court,.

J. R. UPTON,
Clerk to the Society.

APOTHECARIES' HALL,
LONDON, 1878.

For the following description and plans of the Garden, and the Catalogue of the plants it contains, the Editor is mainly indebted to the co-operation of the distinguished Curator, Mr. Thomas Moore.

1878 Present state of the Garden.

The Chelsea Botanic Garden at the present time occupies about four acres of land. The figure of the Garden is that of a trapezium, the four sides not being equal in length, and the inequality is now increased by the addition of a slip of reclaimed land, which has been thrown in with the older portion since the construction of the new river-side embankment. At the eastern end of the side facing the river, the angle formed by the boundaries is an acute one, and at the western end the angle is obtuse, while the side looking on to the river is longer than that running parallel to the Queen's Road at the northern boundary. The old wall still incloses the Garden on three of its sides, but on the embankment front, or southern side, a handsome and substantial iron fence, with entrance-gates, has recently been erected.

The general disposition of the ground is that shown in Plan I. Near the centre stands a marble statue by Rysbrach of Sir Hans Sloane, Bart., the donor of the ground. On the right of the entrance-gate (1), on the plot marked A, is a new plantation of hardy medicinal plants recently made. This is shown more in detail on Plan II., and is planted in serial order after the system of De Candolle. Westward from this, on Plots B and C, as shown in detail on Plan III., is an older plantation of medicinal plants arranged according to Lindley's "Flora Medica." This is not quite so extensive a series as the former, and many of the subjects are duplicates, which is found rather an advantage than otherwise, as from variations of soil and aspect, and in some degree also from treatment, a longer succession of flowers is obtained than would be furnished by a single patch of each plant.

It will be seen that from the entrance in Swan Walk (1) a gravel path crosses the Garden from east to west, and this is crossed by another from north to south, the statue of Sir Hans Sloane standing at the point where the paths intersect. Starting from the same point another path skirts the four sides of the Garden, while a further loop-line traverses the new ground added near the embankment. In general terms, therefore, the portion of the Garden to the right or north of the centre path entering from Swan Walk is devoted to medical plants, and that on the left hand or south of this path, indicated by the plots D, E, F, G, to beds of hardy herbaceous plants arranged in natural orders, a few trees and shrubs, mostly the remains of former arrangements, being interspersed. H is an enclosed yard for storing fuel, soils, and other necessary materials.

There are four glass houses, besides pits and frames devoted to propagation. The reference figure 2 indicates the position of a span-roofed house divided in the interior into stove and greenhouse, and primarily devoted to such medicinal plants as will not grow out-doors.

The lean-to houses 3 and 4 have their contents planted out *à la* Ward, and may be taken to represent on a very small scale the mixed vegetations of climates warmer than our own. The plants are not medical but of general botanical interest, No. 3 containing a considerable number of succulents. The house No. 5 is the cold house already mentioned, also planted *à la* Ward. At 6 and 7 are warm propagating pits, 8 is a cold pit, and 9 indicates the site of cold frames for various cultural uses in rearing young plants to supply the different collections in the Garden. A tank for water-plants (10) is situated at the lower part of the plot E. The plot I is intended for annual flowers, arranged in a serial manner according to their natural orders; 11 is a bed of Yuccas; 12, 12 are mixed borders of hardy flowers on the east side, useful as affording different aspects for some of the same plants which occur in other parts of the Garden; 13 is a similar border on the west side partly filled with shrubby plants; while 14, 14 are grassy slopes running up to the embankment, having a series of mixed ornamental shrubs within the low inner wall, and a holly hedge between the inner wall and the outer fence.

The extreme eastern corner is reserved as a plot for nursery purposes.

Within the area of the yard H stands one of the finest Oriental Planes (*Platanus orientalis*) to be seen in or about London, a tree of noble proportions, which it may be hoped will not materially suffer from its water supply being cut off by the embankment. On plot F. stands the sole remaining Cedar of Lebanon (*Cedrus Libani*), one of the original trees introduced to this country; and on plot G is a handsome, straight-stemmed, evergreen oak (*Quercus ilex*). On the wall skirting Swan Walk, south of the entrance-gate, is an old tree of *Styrax officinalis*, and on the northern side a fine *Wistaria*, while close to the north boundary wall, near bed 11, stands a very good example of the Maidenhair tree (*Salisburia adiantifolia*).

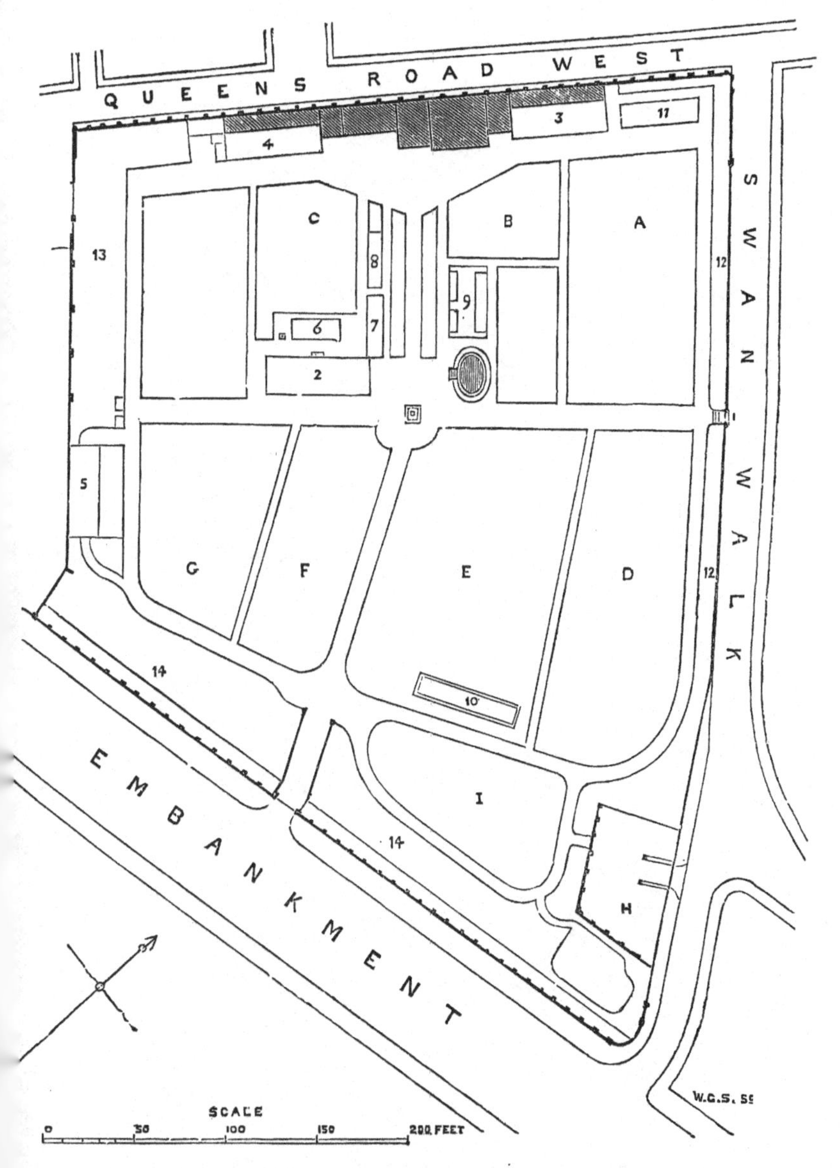

PLAN. I.—Ground Plan of Chelsea Botanic Garden as it is in 1878.

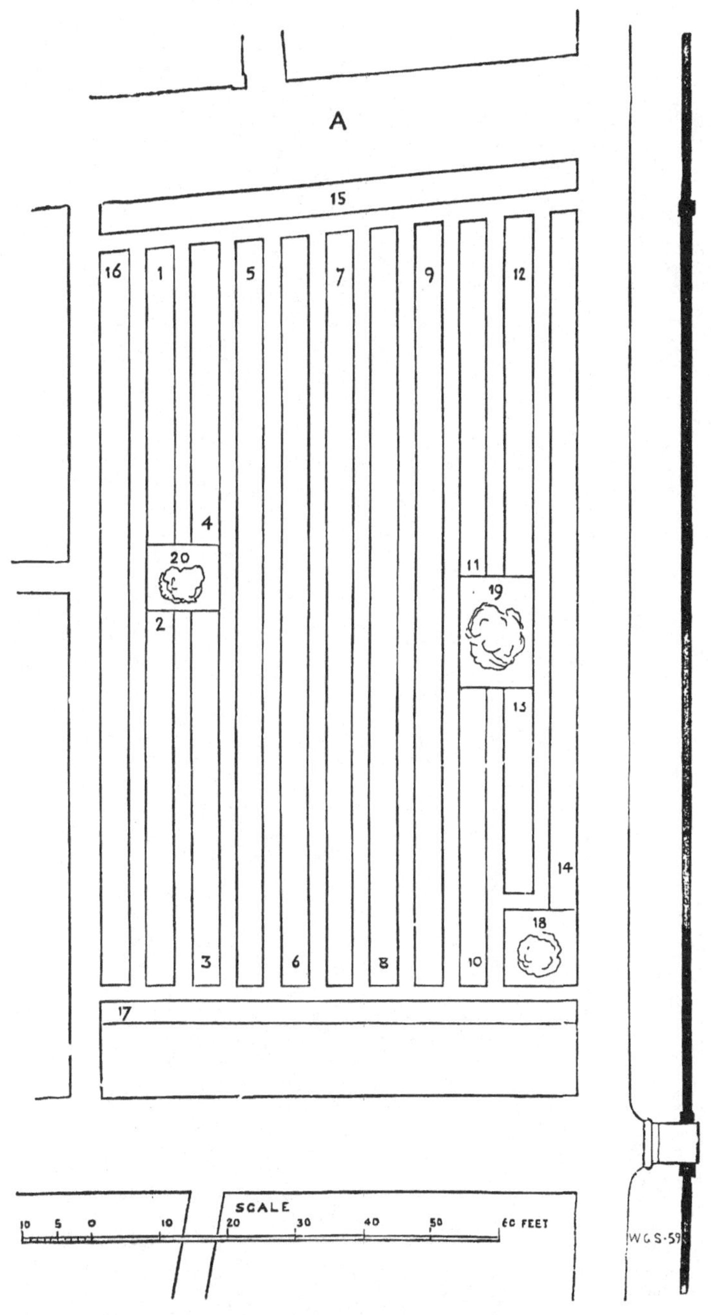

PLAN II.—Plan showing Beds of Hardy Medicinal Plants, arranged after De Candolle on Plot A of Plan I.

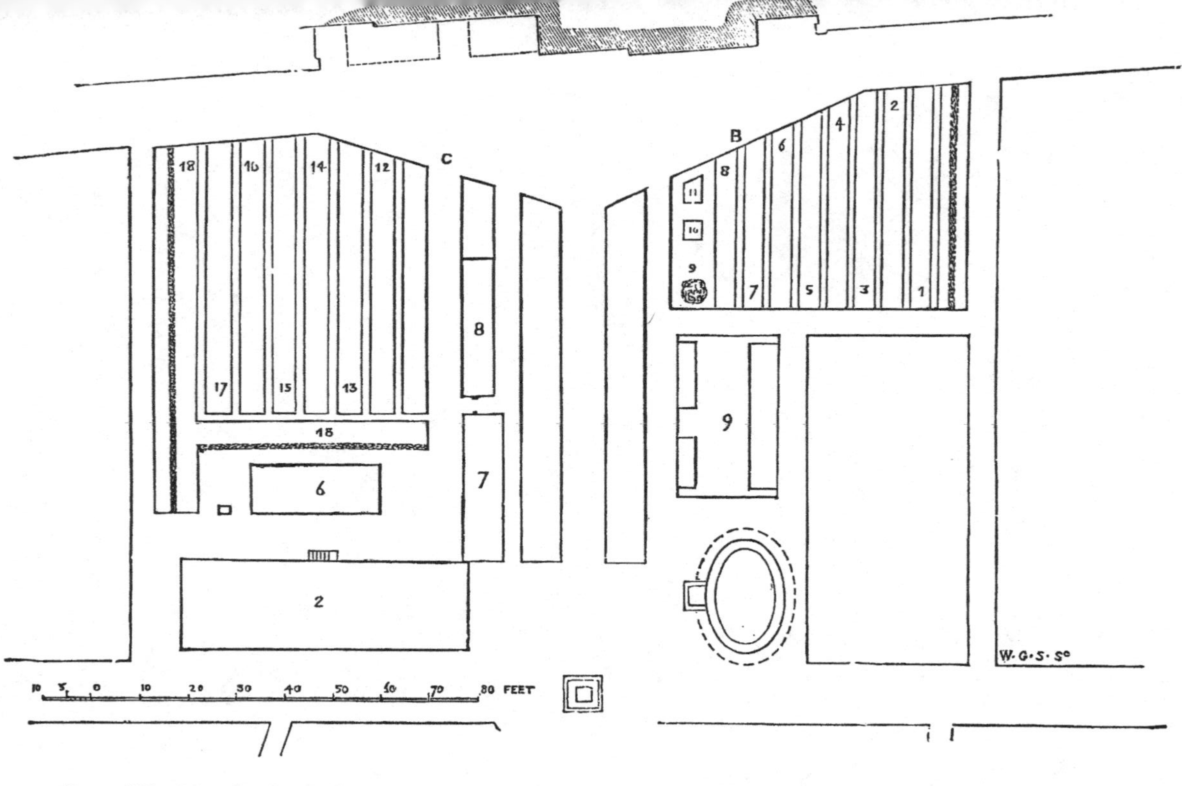

PLAN III.—Plan showing Beds of Hardy Medicinal Plants, arranged after Lindley, on Plots B and C of Plan I.

LIST OF MEDICAL PLANTS IN PLOT A,

AFTER THE ARRANGEMENT OF DE CANDOLLE.

(*See* PLAN II.)

BED 1, commencing at N. end.

Order Ranunculaceæ.

Clematis erecta, *All.*—Central Europe.
Clematis Flammula, *Lin.*—South Europe.
Anemone Pulsatilla, *Lin.*—England.
Ranunculus acris, *Lin.*—England.
Ranunculus Flammula, *Lin.*—England.
Helleborus fœtidus, *Lin.*—England.
Helleborus niger, *Lin.*—Central Europe.
Helleborus orientalis, *Lam.*—Asia Minor, Greece.
Aconitum Napellus, *Lin.*—England.
Aconitum ferox, *Wall.*—Nepal.
Aconitum paniculatum, *Lam.*—Europe.
Aconitum Lycoctonum, *Lin.*—Europe.
Delphinium Consolida, *Lin.*—England.

BED 2.

Delphinium Staphisagria, *Lin.*—South Europe.
Pæonia officinalis, *Retz.*—Europe.
Actæa spicata, *Lin.*—Europe.
Xanthorrhiza apiifolia, *Herit.*—Virginia.
Cimicifuga racemosa, *Bart.*—North America.
Podophyllum peltatum, *Lin.*—North America.

Order Magnoliaceæ.

Liriodendron Tulipifera, *Lin.*—North America.
Magnolia glauca, *Lin.*—North America.

Order Berberidaceæ.

Berberis vulgaris, *Lin.*—England.

Order Papaveraceæ.

Chelidonium majus, *Lin.*—England.
Argemone mexicana, *Lin.*—Mexico.
Papaver Rhœas, *Lin.*—England.
Papaver somniferum, *Lin.*—Levant, South Europe.

Order Cruciferæ.

Isatis tinctoria, *Lin.*—England.
Cochlearia Armoracia, *Lin.*—Europe.
Cochlearia officinalis, *Lin.*—England.

BED 3, S. end.

Cardamine pratensis, *Lin.*—England.
Sinapis alba, *Lin.*—England.
Sinapis nigra, *Lin.*—England.

Order Violaceæ.

Viola odorata, *Lin.*—England.
Viola canina, *Lin.*—England.

Order Polygalaceæ.

Polygala vulgaris, *Lin.*—England.

Order Caryophyllaceæ.

Agrostemma Githago, *Lin.*—England.
Dianthus Caryophyllus, *Lin.*—England.
Saponaria officinalis, *Lin.*—England.
Vaccaria vulgaris, *Host*—Europe.

Order Tamaricaceæ.

Tamarix gallica, *Lin.*—England, Mediterranean region.
Myricaria germanica, *Desv.*—Europe, Caucasus.

Order Hypericaceæ.

Androsæmum officinale, *All.*—England, Greece.

Order Malvaceæ.

Althæa officinalis, *Lin.*—England.
Malva sylvestris, *Lin.*—England.

Order Tiliaceæ.

Tilia europæa, *Lin.*—England.

BED 4.

Order Linaceæ.

Linum usitatissimum, *Lin.*—England.

Order Oxalidaceæ.

Oxalis crenata, *Jacq.*—Mountains of Peru.

Order Geraniaceæ.

Geranium maculatum, *Lin.*—North America.
Geranium Robertianum, *Lin.*—England.
Erodium moschatum, *Willd.*—England.

Order Zygophyllaceæ.

Zygophyllum Fabago, *Lin.*—Syria, Morocco.

Order Rutaceæ.

Ruta graveolens, *Lin.*—South Europe.
Dictamnus albus, *Lin.*—South Europe.

Order Xanthoxylaceæ.

Xanthoxylon fraxineum, *Willd.*—North America.
Ptelea trifoliata, *Lin.*—North America.

Order Aquifoliaceæ.

Ilex Aquifolium, *Lin.*—England.

Order Celastraceæ.

Euonymus europæus, *Lin.*—England.

Order Rhamnaceæ.

Rhamnus catharticus, *Lin.*—England.

BED 5, N. end.

Rhamnus infectorius, *Lin.*—South Europe.

Order Vitaceæ.

Vitis vinifera, *Lin.*—South Asia, Greece, &c.

Order Sapindaceæ.

Æsculus Hippocastanum, *Lin.*—Asia, North India.
Æsculus ohioensis, *Michaux*—North America.

Order Anacardiaceæ.

Pistacia Lentiscus, *Lin.*—South Europe, North Africa, Levant.

Rhus Toxicodendron, *Lin.*—North America.
Rhus Cotinus, *Lin.*—South Europe, Caucasus.
Rhus glabra, *Lin.*—North America.

Order Coriariaceæ.

Coriaria myrtifolia, *Lin.*—Mediterranean region.

Order Leguminosæ.

Baptisia tinctoria, *R. Br.*—North America.
Astragalus Pseudo-Tragacantha, *Pall.*—Eastern Caucasus.
Spartium junceum, *Lin.*—South Europe.
Sarothamnus scoparius, *Wimm.*—England.
Trigonella cœrulea, *Seringe*—South Europe, North Africa.
Trigonella fœnum-græcum, *Lin.*—South of France.
Glycyrrhiza glabra, *Lin.*—South Europe.
Colutea arborescens, *Lin.*—Central and South Europe.
Ervum Ervilia, *Lin.*—South Europe.
Ervum Lens, *Lin.*—Europe.
Faba vulgaris, *Mœnch.*—Caspian region.
Lathyrus tuberosus, *Lin.*—England.
Coronilla Emerus, *Lin.*—Central and South Europe.

Order Drupaceæ.

Amygdalus communis dulcis, *DC.*—Levant, North Africa.
Amygdalus communis amara, *DC.*—Levant, North Africa.
Amygdalus persica, *Lin.*—Persia.
Cerasus virginiana, *Michaux*—North America.
Cerasus Laurocerasus, *Loisel.*—Trebizond.
Prunus spinosa, *Lin.*—England.
Prunus Armeniaca, *Lin.*—Armenia.

Order Pomaceæ.

Pyrus Aucuparia, *Gærtn.*—England.
Mespilus germanica, *Lin.*—England.
Cydonia vulgaris, *Pers.*—South Europe.

BED 6, S. end.

Order Rosaceæ.

Rosa centifolia, *Lin.*—East Caucasus.
Rosa gallica, *Lin.*—Central and South Europe.
Rosa canina, *Lin.*—England.
Rubus Idæus, *Lin.*—England.
Rubus villosus, *Ait.*—North America.
Potentilla reptans, *Lin.*—England.
Potentilla Tormentilla, *Nestl.*—England.
Agrimonia Eupatoria, *Lin.*—England.
Geum urbanum, *Lin.*—England.
Geum rivale, *Lin.*—England.
Geum canadense, *Murr.*—North America.
Spiræa Ulmaria, *Lin.*—England.
Poterium Sanguisorba, *Lin.*—England.

Order Calycanthaceæ.

Calycanthus floridus, *Lin.*—Carolina.

Order Myrtaceæ.

Punica Granatum, *Lin.*—Syria, North Africa, &c.

Order Lythraceæ.

Lythrum Salicaria, *Lin.*—England.

Order Cucurbitaceæ.

Bryonia dioica, *Jacq.*—England.
Cucurbita aurantia, *Willd.*—Patria ignota.
Ecballium Elaterium, *Kich.*—South Europe.

Order Datiscaceæ.

Datisca cannabina, *Lin.*—Asia Minor, Persia, North India.

Order Crassulaceæ.

Sedum acre, *Lin.*—England.
Sedum Telephium, *Lin.*—England.
Sempervivum tectorum, *Lin.*—England.

Order Saxifragaceæ.

Heuchera americana, *Lin.*—North America.

Order Grossularieæ.

Ribes nigrum, *Lin.*—England.

Order Umbelliferæ.

Eryngium maritimum, *Lin.*—England.
Cicuta virosa, *Lin.*—England.
Apium graveolens, *Lin.*—England.
Petroselinum sativum, *Hoffm.*—England.
Sium Sisarum, *Lin.*—China, Japan, Mongolia, &c.
Pimpinella Saxifraga, *Lin.*—England.
Carum Carui, *Lin.*—England.

Bed 7, N. end.

Œnanthe crocata, *Lin.*—England.
Æthusa Cynapium, *Lin.*—England.
Fœniculum vulgare, *Gærtn.*—England.
Meum Athamanticum, *Jacq.*—England.
Archangelica officinalis, *Hoffm.*—Europe.
Opopanax Chironium, *Koch*—South Europe.
Ferula persica, *Willd.*—Persia.
Ferula Ferulago, *Lin.*—Mediterranean region.
Ferula tingitana, *Lin.*—North Africa, Tangiers.
Narthex Asafœtida, *Falconer*—North-west India.
Imperatoria Ostruthium, *Lin.*—England.
Anethum graveolens, *Lin.*—Central Europe, Egypt, &c.
Pastinaca sativa, *Lin.*—England.
Peucedanum officinale, *Lin.*—England.
Daucus Carota, *Lin.*—England.
Anthriscus Cerefolium, *Hoffm.*—South Europe.
Conium maculatum, *Lin.*—England.
Smyrnium Olusatrum, *Lin.*—England.
Coriandrum sativum, *Lin.*—England.

Order Araliaceæ.

Gunnera chilensis, *Lam.*—Chili, Peru.
Hedera Helix, *Lin.*—England.
Aralia nudicaulis, *Lin.*—North America.

Order Cornaceæ (Rubiaceæ).

Cornus mascula, *Lin.*—South Europe.
Cornus circinata, *Herit.*—North America.
Cornus sericea, *Herit.*—North America.
Cornus florida, *Lin.*—North America.

Order Caprifoliaceæ.

Sambucus nigra, *Lin.*—England.

Order Galiaceæ (Rubiaceæ).

Asperula odorata, *Lin.*—England.
Rubia tinctorum, *Lin.*—South Europe, Levant.
Galium Aparine, *Lin.*—England.

Order Valerianaceæ.

Valeriana celtica, *Lin.*—Alps of Europe (in frames).
Valeriana officinalis, *Lin.*—England.

BED 8, S. end.

Valeriana Phu, *Lin.*—Central Europe.

Order Compositæ, § Tubulifloræ.

Tussilago Farfara, *Lin.*—England.
Inula Helenium, *Lin.*—England.
Anthemis nobilis, *Lin.*—England.
Achillea Millefolium, *Lin.*—England.
Achillea nobilis, *Lin.*—Central and South Europe.
Ptarmica vulgaris, *DC.*—England.
Pyrethrum Tanacetum, *DC.*—South Europe.
Pyrethrum Parthenium, *Smith*—England.
Artemisia Abrotanum, *Lin.*—South Europe.
Artemisia Absinthium, *Lin.*—England.
Artemisia pontica, *Lin.*—Europe, Central Asia.
Artemisia Santonica, *Lin.*—Tartary, Siberia.
Artemisia Dracunculus, *Lin.*—South Europe, Siberia.
Tanacetum vulgare, *Lin.*—England.
Doronicum Pardalianches, *Lin.*—England.
Arctium majus, *Schkuhr*—England.
Cnicus benedictus, *Gærtn.*—South Europe, Persia.
Carthamus tinctorius, *Lin.*—India, Egypt.
Cynara Scolymus, *Lin.*—South Europe.

Order Compositæ, § Liguliforæ.

Cichorium Endivia, *Lin.*—India, Asia Minor, Greece.
Cichorium Intybus, *Lin.*—England.
Tragopogon porrifolius, *Lin.*—England.

Scorzonera hispanica, *Lin.*—Europe, North Asia.
Lactuca sativa, *Lin.*—Europe.
Lactuca virosa, *Lin.*—England.
Taraxacum Dens leonis, *Desf.*—England.

Order Lobeliaceæ.
Lobelia cardinalis, *Lin.*—North America.
Lobelia syphilitica, *Lin.*—North America.
Lobelia inflata, *Lin.*—North America.

Order Ericaceæ.
Arbutus Unedo, *Lin.*—South Europe, Ireland.
Arctostaphylos Uva-ursi, *Spr.*—England.

BED 9, N. end.

Gaultheria procumbens, *Lin.*—North America.
Kalmia latifolia, *Lin.*—North America.
Ledum latifolium, *Ait.*—North America.
Rhododendron ferrugineum, *Lin.*—Switzerland.
Azalea pontica, *Lin.*—Asia Minor, Caucasus.

Order Ebenaceæ.
Diospyros virginiana, *Lin.*—North America.

Order Primulaceæ.
Anagallis arvensis, *Lin.*—England.
Cyclamen europæum, *Lin.*—South Europe, England.
Primula veris, *Lin* —England.

Order Jasminaceæ.
Jasminum officinale, *Lin.*—China, South Europe.

Order Apocynaceæ.
Apocynum androsæmifolium, *Lin.*—North America.
Apocynum cannabinum, *Lin.*—North America.

Order Asclepiadaceæ.
Asclepias tuberosa, *Lin.*—North America.
Vincetoxicum officinale, *Mœnch*—Europe.

Order Oleaceæ.
Fraxinus excelsior, *Lin.*—England.
Fraxinus Ornus, *Lin.*—South Europe.
Olea europæa, *Lin.*—South Europe, Barbary, Levant.
Syringa vulgaris, *Lin.*—East Europe, Persia.

Order Gentianaceæ.

Menyanthes trifoliata, *Lin.*—England.
Erythræa Centaurium, *Pers.*—England.
Gentiana lutea, *Lin.*—Alps of Europe.

Order Convolvulaceæ.

Calystegia sepium, *R. Br.*—England.
Convolvulus arvensis, *Lin.*—England.
Convolvulus Scammonia, *Lin.*—Greece, Levant.
Pharbitis Nil, *Choisy*—India, Tropics generally.

Order Solanaceæ.

Nicotiana Tabacum, *Lin.*—Tropical America.
Datura Stramonium, *Lin.*—England.
Hyoscyamus niger, *Lin.*—England.
Atropa Belladonna, *Lin.*—England.
Capsicum annuum, *Lin.*—India, South America, Mexico.
Solanum Dulcamara, *Lin.*—England.
Solanum nigrum, *Lin.*—England.

BED 10, S. end.

Lycopersicum esculentum, *Mill.*—South America, Peru.

Order Boraginaceæ.

Anchusa tinctoria, *Lin.*—South Europe.
Borago officinalis, *Lin.*—England.

Order Labiatæ.

Ocimum Basilicum, *Lin.*—India, Western Asia.
Lavandula Spica, *DC.*—Mediterranean region, Algiers.
Lavandula vera, *DC.*—Mediterranean region, Morocco.
Mentha piperita, *Lin.*—England.
Mentha viridis, *Lin.*—England.
Mentha Pulegium, *Lin.*—England.
Mentha aquatica, *Lin.*—England.
Salvia officinalis, *Lin.*—South Europe.
Rosmarinus officinalis, *Lin.*—South Europe, Asia Minor, Algiers.

BED 11.

Origanum Majorana, *Lin.*—Central Asia, Arabia, North Africa.

Origanum vulgare, *Lin.*—England.
Satureia hortensis, *Lin.*—South Europe.
Thymus Serpyllum, *Lin.*—England.
Hyssopus officinalis, *Lin.*—South Europe, Central Asia.
Melissa officinalis, *Lin.*—South Europe, England.
Nepeta Cataria, *Lin.*—England.
Nepeta Glechoma, *Benth.*—England.
Marrubium vulgare, *Lin.*—England.
Teucrium Chamædrys, *Lin.*—England.

Order Verbenaceæ.

Verbena officinalis, *Lin.*—England.

Order Scrophulariaceæ.

Verbascum Thapsus, *Lin.*—England.
Linaria vulgaris, *Mill.*—England.
Scrophularia nodosa, *Lin.*—England.
Gratiola officinalis, *Lin.*—Europe.

Bed 12, N. end.

Digitalis purpurea, *Lin.*—England.

Order Plumbaginaceæ.

Armeria vulgaris, *Willd.*—England.
Plumbago europæa, *Lin.*—South Europe, Asia Minor.

Order Plantaginaceæ.

Plantago Psyllium, *Lin.*—South Europe.
Plantago Cynops, *Lin.*—Europe.

Order Chenopodiaceæ.

Chenopodium anthelminticum, *Lin.*—North America, Brazil.
Chenopodium Vulvaria, *Lin.*—England.
Chenopodium Botrys, *Lin.*—South Europe, North Africa.

Order Nyctaginaceæ.

Mirabilis Jalapa, *Lin.*—Peru, Mexico.
Mirabilis longiflora, *Lin.*—Mexico.

Order Phytolaccaceæ.

Phytolacca decandra, *Lin.*—North America.

Order Polygonaceæ.

Rheum undulatum, *Lin.*—Siberia.
Rheum palmatum, *Lin.*—Chinese Tartary.
Rheum Emodi, *Wall.*—Himalaya.
Rheum officinale, *Baillon*—Eastern Thibet, Western China.

BED 13.

Rumex Hydrolapathum, *Huds.*—England.
Rumex alpinus, *Lin.*—England.
Rumex Acetosa, *Lin.*—England.
Polygonum Bistorta, *Lin.*—England.

Order Thymelaceæ.

Daphne Mezereum, *Lin.*—England.
Daphne pontica, *Lin.*—Asia Minor.
Daphne Laureola, *Lin.*—England.

Order Elæagnaceæ.

Elæagnus angustifolia, *Lin.*—South Europe, Central Asia.

BED 14, S. end.

Order Lauraceæ.

Sassafras officinale, *Nees*—North America.
Laurus nobilis, *Lin.*—South Europe, Asia Minor.
Benzoin odoriferum, *Nees*—North America.

Order Buxaceæ.

Buxus sempervirens, *Lin.*—England.

Order Euphorbiaceæ.

Ricinus communis, *Lin.*—India.
Mercurialis annua, *Lin.*—England.
Mercurialis perennis, *Lin.*—England.
Euphorbia Cyparissias, *Lin.*—England.
Euphorbia Gerardiana, *Jacq.*—Europe.

Euphorbia hyberna, *Lin.*—England.
Euphorbia palustris, *Lin.*—Europe.
Euphorbia Lathyris, *Lin.*—England.
Euphorbia Peplus, *Lin.*—England.
Euphorbia helioscopia, *Lin.*—England.

Order Empetraceæ.

Empetrum nigrum, *Lin.*—England.

Order Corylaceæ.

Quercus pedunculata, *Willd.*—England.
Quercus sessiliflora, *Smith*—England.
Quercus lusitanica (infectoria), *Webb*—Mediterranean region.

Order Juglandaceæ.

Juglans regia, *Lin.*—Asia Minor, Levant, &c.

Order Moraceæ.

Ficus Carica, *Lin.*—South Europe, Asia Minor.
Morus alba, *Lin.*—Asia.
Morus nigra, *Lin.*—South Europe, Persia.

Order Cannabinaceæ.

Humulus Lupulus, *Lin.*—England.
Cannabis sativa, *Lin.*—India, Persia.

Order Urticaceæ.

Parietaria officinalis, *Lin.*—England.
Urtica dioica, *Lin.*—England.

BED 15, E. end.

Order Ulmaceæ.

Ulmus montana, *Smith*—England.
Celtis australis, *Lin.*—South Europe, North Africa.

Order Aristolochiaceæ.

Aristolochia Clematitis, *Lin.*—Europe, England.
Asarum canadense, *Lin.*—North America.
Asarum europæum, *Lin.*—England.

Order Saururaceæ.

Saururus cernuus, *Lin.*—North America.

Order Myricaceæ.

Comptonia aspleniifolia, *Gærtn.*—North America.
Myrica Gale, *Lin.*—England.
Myrica cerifera, *Lin.*—North America.

Order Betulaceæ.

Alnus glutinosa, *Willd.*—England.
Betula nigra, *Lin.*—North America.
Betula alba, *Lin.*—England.

Order Salicaceæ.

Populus balsamifera, *Lin.*—North America.
Populus nigra, *Lin.*—England.
Salix pentandra, *Lin.*—England.
Salix fragilis, *Lin.*—England.
Salix Russelliana, *Smith*—England.
Salix purpurea, *Lin.*—England.

Order Platanaceæ.

Liquidambar styraciflua, *Lin.*—North America.

Order Coniferæ, § Taxineæ.

Taxus baccata, *Lin.*—England.

BED 16, N. end.

Order Coniferæ, § Abietineæ.

Picea balsamea, *Loud.*—North America.
Picea pectinata, *Loud.*—Central and South Europe.
Abies nigra, *Michaux*—North America.
Abies excelsa, *DC.*—North Europe.
Pinus Pinaster, *Soland.*—South Europe.
Pinus sylvestris, *Lin.*—England.
Pinus Pumilio, *Hænke*—Central Europe.
Larix europæa, *DC.*—European Alps, Siberia.
Juniperus communis, *Lin.*—England.
Juniperus Sabina, *Lin.*—Central Europe.
Thuja occidentalis, *Lin.*—North America.

Order Dioscoreaceæ.

Tamus communis, *Lin.*—England.

Order Smilaceæ.

Smilax aspera, *Lin.*—South Europe, Levant.

Order Marantaceæ.

Canna edulis, *Ker*—Peru.

Order Hæmodoraceæ.

Aletris farinosa, *Lin.*—North America.

Order Amaryllidaceæ.

Leucoium æstivum, *Lin.*—England.
Narcissus Pseudo-Narcissus, *Lin.*—England.
Narcissus Tazetta, *Lin.*—South Europe, North Africa.
Pancratium maritimum, *Lin.*—Mediterranean region.

Order Iridaceæ.

Crocus sativus, *Lin.*—Greece, Asia Minor.
Iris fœtidissima, *Lin.*—England.
Iris florentina, *Lin.*—South Europe.
Iris germanica, *Lin.*—Europe.
Iris Pseudacorus, *Lin.*—England.

Order Liliaceæ, § Tulipeæ.

Fritillaria imperialis, *Lin.*—European Turkey, Persia.
Lilium candidum, *Lin.*—South Europe, Asia Minor.

Order Liliaceæ, § Scilleæ.

Allium ascalonicum, *Lin.*—Asia Minor.
Allium Cepa, *Lin.*—Mountains of Caucasus.
Allium fistulosum, *Lin.*—Siberia, Altaic Alps.
Allium Porrum, *Lin.*—South Europe, Egypt.
Allium sativum, *Lin.*—Sicily, Egypt.
Allium Scorodoprasum, *Lin.*—England.
Allium Schœnoprasum, *Lin.*—England.

BED 17, W. end.

Order Liliaceæ, § Asparageæ.

Ruscus aculeatus, *Lin.*—England.
Polygonatum officinale, *All.*—England.
Asparagus officinalis, *Lin.*—England.

Order Melanthaceæ.

Colchicum variegatum, *Lin.*—South Europe, Levant.
Colchicum autumnale, *Lin.*—England.
Veratrum album, *Lin.*—Europe.
Veratrum nigrum, *Lin.*—Central and South Europe, Siberia.

Order Orontiaceæ.

Acorus Calamus, *Lin.*—England.

Order Araceæ.

Arum maculatum, *Lin.*—England.
Dracunculus vulgaris, *Schott*—South Europe.

Order Graminaceæ, § Phalareæ.

Zea Mays, *Lin.*—Paraguay.

Order Graminaceæ, § Paniceæ.

Panicum miliaceum, *Lin.*—India.
Setaria italica, *Kunth*—Europe, India, New Holland.

Order Graminaceæ, § Aveneæ.

Avena sativa, *Lin.*—Mesopotamia.

Order Graminaceæ, § Hordeæ.

Lolium temulentum, *Lin.*—England.
Triticum repens, *Lin.*—England.
Triticum vulgare, *Vill.*—Central Asia.
Secale cereale, *Lin.*—Caucasico-Caspian Desert.
Hordeum distichum, *Lin.*—Mesopotamia, Tartary.

Order Cyperaceæ.

Carex hirta, *Lin.*—England.
Carex arenaria, *Lin.*—England.

LIST OF MEDICAL PLANTS IN PLOTS B AND C,

AFTER THE ARRANGEMENT OF LINDLEY.

(*See* PLAN III.)

BED 1, commencing at S. end.

Order Ranunculaceæ.

Clematis Flammula, *Lin.*—South Europe.
Clematis erecta, *All.*—Central Europe.
Clematis Vitalba, *Lin.*—England.
Anemone coronaria, *Lin.*—South Europe.
Adonis vernalis, *Lin.*—Europe.
Ranunculus acris, *Lin.*—England.
Ranunculus bulbosus, *Lin.*—England.
Helleborus niger, *Lin.*—Central Europe.
Helleborus fœtidus, *Lin.*—England.
Helleborus viridis, *Lin.*—England.
Delphinium Staphisagria, *Lin.*—South Europe.
Delphinium Consolida, *Lin.*—England.
Aconitum Napellus, *Lin.*—England.
Aconitum paniculatum, *Lam.*—Europe.
Pæonia officinalis, *Retz.*—Europe.

BED 2, N. end.

Podophyllum peltatum, *Lin.*—North America.

Order Papaveraceæ

Sanguinaria canadensis, *Lin.*—North America.
Papaver somniferum, *Lin.*—Levant, South Europe.
Papaver Rhœas, *Lin.*—England.
Chelidonium majus, *Lin.*—England.

Order Umbelliferæ.

Cicuta virosa, *Lin.*—England.
Apium graveolens, *Lin.*—England.
Petroselinum sativum, *Hoffm.*—England.
Carum Carui, *Lin.*—England.
Œnanthe crocata, *Lin.*—England.
Archangelica officinalis, *Hoffm.*—Europe.
Æthusa Cynapium, *Lin.*—England (on border by adjacent Holly hedge).
Fœniculum vulgare, *Gærtn.*—England.
Eryngium campestre, *Lin.*—England.

Bed 3, S. end.

Dorema Asafœtida, *Hort. Edin.*—Persia.
Opopanax Chironium, *Koch.*—South Europe.
Ferula persica, *Willd.*—Persia.
Ferula tingitana, *Lin.*—North Africa.
Narthex Asafœtida, *Falc.*—North-West India.
Imperatoria Ostruthium, *Lin.*—England.
Anethum graveolens, *Lin.*—Central Europe.
Heracleum Sphondylium, *Lin.*—England.
Daucus Carota, *Lin.*—England.
Anthriscus sylvestris, *Hoffm.*—England.
Conium maculatum, *Lin.*—England.
Smyrnium Olusatrum, *Lin.*—England.
Coriandrum sativum, *Lin.*—England.

Bed 4, N. end.

Order Berberidaceæ.

Berberis vulgaris, *Lin.*—England.

Order Vitaceæ.

Vitis vinifera, *Lin.*—South Asia, Greece, &c.

Order Myrtaceæ.

Punica Granatum, *Lin.*—Syria, North Africa, &c.

Order Cucurbitaceæ.

Ecballium Elaterium, *Rich.*—South Europe.

Order Cruciferæ.

Cochlearia Armoracia, *Lin.*—Europe.
Cochlearia officinalis, *Lin.*—England.
Cardamine pratensis, *Lin.*—England.
Sinapis alba, *Lin.*—England.
Sinapis nigra, *Lin.*—England.
Crambe maritima, *Lin.*—England.

Order Linaceæ.

Linum usitatissimum, *Lin.*—England.

BED 5, S. end.

Order Malvaceæ.

Malva sylvestris, *Lin.*—England.
Althæa officinalis, *Lin.*—England.

Order Lythraceæ.

Lythrum Salicaria, *Lin.*—England.

Order Rhamnaceæ.

Rhamnus catharticus, *Lin.*—England (on Plot E)

Order Buxaceæ.

Buxus sempervirens, *Lin.*—England.

Order Euphorbiaceæ.

Ricinus communis, *Lin.*—India.
Mercurialis perennis, *Lin.*—England.
Euphorbia Lathyris, *Lin.*—England.
Euphorbia Cyparissias, *Lin.*—England.

Order Tamaricaceæ.

Tamarix gallica, *Lin.*—England, Mediterranean.

Order Rutaceæ.

Ruta graveolens, *Lin.*—South Europe.
Dictamnus albus, *Lin.*—South Europe.

BED 6, N. end.

Order Geraniaceæ.

Geranium maculatum *Lin.*—North America.
Geranium Robertianum, *Lin.*—England.

Order Oxalidaceæ.

Oxalis crenata, *Jacq.*—Mountains of Peru.

Order Coriariaceæ.

Coriaria myrtifolia, *Lin.*—Mediterranean region.

Order Rosaceæ.

Potentilla Tormentilla, *Nestl.*—England.
Geum rivale, *Lin.*—England.
Geum urbanum, *Lin.*—England.
Agrimonia Eupatoria, *Lin.*—England.
Rubus villosus, *Ait.*—North America.
Rosa canina, *Lin.*—England.

BED 7, S. end.

Rosa centifolia, *Lin.*—East Caucasus.
Rosa gallica, *Lin.*—Central and South Europe.

Order Rosaceæ, § Pruneæ.

Amygdalus communis, *Lin.*—Levant, North Africa.
Cerasus Laurocerasus, *Loisel.*—Trebizond.
Prunus spinosa, *Lin.*—England.
Prunus domestica, *Lin.*—England.

Order Rosaceæ, § Pomeæ.

Pyrus Aucuparia, *Gærtn.*—England.
Cydonia vulgaris, *Pers.*—South Europe.

Order Leguminosæ.

Cytisus Laburnum, *Lin.*—Alps of Europe.

BED 8, N. end.

Sarothamnus sccparius, *Wimm.*—England.
Glycyrrhiza glabra, *Lin.*—South Europe.
Colutea arborescens, *Lin.*—Central and South Europe.
Coronilla Emerus, *Lin.*—Central and South Europe
Coronilla varia, *Lin.*—South Europe.

Order Crassulaceæ.

Sempervivum tectorum, *Lin.*—England.
Sedum acre, *Lin.*—England.

Order Anacardiaceæ.

Rhus Toxicodendron, *Lin.*—North America.

Pistacia Terebinthus, *Lin.*—South Europe, North Africa, Levant, (on Plot F).

BED 9.

Order Moraceæ.

Morus nigra, *Lin.*—South Europe, Persia.

BED 10.

Order Cannabinaceæ.

Humulus Lupulus, *Lin.*—England.

BED 11.

Cannabis sativa, *Lin.*—India, Persia.

BED 12, N. end.

Order Oleaceæ.

Olea europæa, *Lin.*—South Europe, Barbary, Levant.

Order Platanaceæ.

Liquidambar styraciflua, *Lin.*—North America.

Order Thymelaceæ.

Daphne Mezereum, *Lin.*—England.

Order Lauraceæ.

Lindera (Laurus) Benzoin, *Meissn.*—North America.

Sassafras officinale, *Nees.*—North America.

Laurus nobilis, *Lin.*—South Europe, Asia Minor.

Order Aristolochiaceæ.

Aristolochia Serpentaria, *Lin.*—North America (in greenhouse).

Asarum europæum, *Lin.*—England.

Asarum canadense, *Lin.*—North America.

Order Polygonaceæ.

Rheum palmatum *var.* tanghuticum, *Regel.*—Mongolia, Kansu.

Rheum undulatum, *Lin.*—Siberia.

Rumex Acetosa, *Lin.*—England.

Order Ericaceæ.
Azalea pontica, *Lin.*—Asia Minor, Caucasus.
Gaultheria procumbens, *Lin.*—North America.
Arctostaphylos Uva-ursi, *Spr.*—England.

Order Styraceæ.
Styrax officinalis, *Lin.*—Greece, Asia Minor (on east wall).

BED 13, S. end.

Order Convolvulaceæ.
Convolvulus Scammonia, *Lin.*—Greece, Levant.
Exogonium Purga, *Benth.*—Mexico.

Order Lobeliaceæ.
Lobelia cardinalis, *Lin.*—North America.
Lobelia syphilitica, *Lin.*—North America.
Lobelia inflata, *Lin.*—North America.

Order Caprifoliaceæ.
Sambucus Ebulus, *Lin.*—England.
Sambucus nigra, *Lin.*—England.

Order Rubiaceæ.
Rubia tinctorum, *Lin.*—South Europe, Levant.

Order Compositæ, § Senecionideæ.
Tussilago Farfara, *Lin.*—England.

Order Compositæ, § Inuloideæ.
Inula Helenium, *Lin.*—England.

Order Compositæ, § Anthemideæ.
Tanacetum vulgare, *Lin.*—England.
Pyrethrum Parthenium, *Smith*—England.
Artemisia Absinthium, *Lin.*—England.
Artemisia Abrotanum, *Lin.*—South Europe.
Artemisia Dracunculus, *Lin.*—South Europe, Siberia.
Anthemis nobilis, *Lin.*—England.

BED 14, N. end.

Order Compositæ, § Cichoriaceæ.
Lactuca sativa, *Lin.*—Europe.
Lactuca virosa, *Lin.*—England.

Taraxacum Dens-leonis, *Desf.*—England.
Cichorium Intybus, *Lin.*—England.

Order Valerianaceæ.

Valeriana officinalis, *Lin.*—England.

Order Labiatæ.

Lavandula vera, *DC.*—Mediterranean region, Morocco.
Mentha viridis, *Lin.*—England.
Mentha piperita, *Lin.*—England.
Mentha Pulegium, *Lin.*—England.
Salvia officinalis, *Lin.*—South Europe.
Rosmarinus officinalis, *Lin.*—South Europe, Asia Minor, Algiers.
Origanum vulgare, *Lin.*—England.
Hyssopus officinalis, *Lin.*—South Europe, Central Asia.
Melissa officinalis, *Lin.*—South Europe, England.
Nepeta Cataria, *Lin.*—England.
Nepeta Glechoma, *Benth.*—England.

BED 15, S. end.

Leonurus Cardiaca, *Lin.*—England.
Stachys Betonica, *Benth*—England.
Marrubium vulgare, *Lin.*—England.

Order Scrophulariaceæ.

Digitalis purpurea, *Lin.*—England.
Scrophularia nodosa, *Lin.*—England.
Linaria vulgaris, *Mill.*—England.
Gratiola officinalis, *Lin.*—Europe.

Order Solanaceæ.

Hyoscyamus niger, *Lin.*—England.
Atropa Belladonna, *Lin.*—England.
Capsicum annuum, *Lin.*—India, South America, Mexico.
Datura Stramonium, *Lin.*—England.
Solanum Dulcamara, *Lin.*—England.
Nicotiana Tabacum, *Lin.*—Tropical America.

Order Asclepiadaceæ.

Vincetoxicum officinale, *Mœnch.*—Europe.

Order Gentianaceæ.

Gentiana lutea, *Lin.*—Alps of Europe.

Menyanthes trifoliata, *Lin.*—England (in Tank No. 10).

Order Oleaceæ.

Olea europæa, *Lin.*—South Europe, Barbary, Levant. (See Bed 12.)

Bed 16, N. end.

Order Coniferæ.

Pinus sylvestris, *Lin.*—England.

Abies balsamea, *Mill.*—North America.

Abies excelsa, *DC*—North Europe.

Larix europæa, *DC.*—European Alps, Siberia.

Juniperus communis, *Lin.*—England.

Juniperus Sabina, *Lin.*—Central Europe.

Order Taxaceæ.

Taxus baccata, *Lin.*—England.

Order Amaryllidaceæ.

Narcissus poeticus, *Lin.*—England.

Pancratium maritimum, *Lin.*—Mediterranean region.

Order Iridaceæ.

Iris Pseudacorus, *Lin.*—England.

Iris florentina, *Lin.*—South Europe.

Crocus sativus, *Lin.*—Greece, England.

Order Melanthaceæ.

Veratrum album, *Lin.*—Europe, Caucasus.

Veratrum viride, *Ait.*—North America.

Veratrum nigrum, *Lin.*—Central and South Europe, Siberia.

Colchicum autumnale, *Lin.*—England.

Bed 17, N. end.

Order Liliaceæ.

Urginea Scilla, *Steinh.*—South Europe, Mediterranean region (in greenhouse).

Allium Porrum, *Lin.*—South Europe, Egypt.

Allium Cepa, *Lin.*—Caucasian Mountains.

Allium sativum, *Lin.*—Sicily, Egypt.

Order Smilaceæ.

Smilax aspera, *Lin.*—South Europe, Asia Minor.
Smilax tamnoides, *Lin.*—North America.

Order Orontiaceæ.

Acorus Calamus, *Lin.*—England.

Order Graminaceæ.

Triticum vulgare, *Vill.*—Central Asia.
Hordeum distichum, *Lin.*—Mesopotamia, Tartary.
Secale cereale, *Lin.*—Caucasico-Caspian Desert.
Avena sativa, *Lin.*—Mesopotamia.

Order Cyperaceæ.

Carex arenaria, *Lin.*—England.
Carex hirta. *Lin.*—England.
Cyperus longus, *Lin.*—England.

Order Filices.

Pteris aquilina, *Lin.*—England.
Lastrea Filix-mas, *Presl.*—England.
Osmunda regalis, *Lin.*—England (in No. 5 house).

Bed 18, Supplementary, commencing at E. end.

Order Polygalaceæ.

Polygala Chamæbuxus, *Lin.*—Europe.

Order Rosaceæ.

Spiræa Ulmaria, *Lin.*—England.
Spiræa Filipendula, *Lin.*—England.

Order Leguminosæ.

Baptisia tinctoria, *R. Br.*—North America.
Lathyrus tuberosus, *Lin.*—England.

Order Aristolochiaceæ.

Aristolochia Clematitis, *Lin.*—Europe, England.

Order Polygonaceæ.

Polygonum amphibium, *Lin.*—England.

Order Primulaceæ.
Primula veris, *Lin.*—England.

Order Coriariaceæ.
Coriaria myrtifolia, *Lin.*—Mediterranean region.

Order Rubiaceæ (Galiaceæ.)
Asperula odorata, *Lin.*—England.

Order Compositæ, § Asteroideæ.
Solidago odora, *Ait.*—North America.

Order Compositæ, § Anthemideæ.
Ptarmica vulgaris, *DC.*—England.

Order Compositæ, § Calendulaceæ.
Calendula officinalis, *Lin.*—South Europe.

Order Compositæ, § Cynaroideæ.
Centaurea Centaurium, *Lin.*—Italy.
Centaurea Jacea, *Lin.*—Central Europe.
Silybum marianum, *Gærtn.*—England.
Cnicus benedictus, *Gærtn.*—South Europe.
Cynara Scolymus, *Lin.*—South Europe.

Order Plantaginaceæ.
Plantago Cynops, *Lin.*—Europe.

Order Plumbaginaceæ.
Plumbago europæa, *Lin.*—South Europe, Asia Minor.

Order Boraginaceæ.
Symphytum officinale, *Lin.*—England.
Borago officinalis, *Lin.*—England.

Order Labiatæ.
Lycopus europæus, *Lin.*—England.

Order Solanaceæ.
Solanum nigrum, *Lin.*—England.
Solanum guineense, *Lam.*—Guinea, Brazil.

Order Amaryllidaceæ.
Oporanthus luteus, *Herb.*—South Europe.
Narcissus Pseudo-Narcissus, *Lin.*—England.

Order Iridaceæ.

Iris versicolor, *Lin.*—North America.

Order Liliaceæ.

Allium Schœnoprasum, *Lin.*—England.

Order Graminaceæ.

Lolium temulentum, *Lin.*—England.
Setaria italica, *Kunth.*—Europe, India.

Order Cyperaceæ.

Cyperus esculentus, *Lin.*—South Europe, Levant.

LIST OF NATURAL ORDERS OF HERBACEOUS PLANTS ON PLOTS D, E, F, G.

(*See* PLAN I.)

THE plants grown on the South side of the Central path, on Plots D, E, F, G, are, as already stated, chiefly herbaceous perennials grouped in beds, each bed containing usually one Natural Order. These Orders are not in any regular sequence, having originally been disposed to fit the open spaces on the ground. They will, however, be readily found on the several Plots, as follows:

Natural Orders on Plot D.

Scrophulariaceæ.
Compositæ.
Dipsaceæ.
Valerianaceæ.
Campanulaceæ.
Lobeliaceæ.
Caprifoliaceæ (shrubby plants).
Galiaceæ (Rubiaceæ).
Umbelliferæ.
Araliaceæ.
Aristolochiaceæ.
Ribesiaceæ (shrubby plants).
Philadelphaceæ (shrubby plants).
Onagraceæ.

The most interesting trees on this Plot are *Kölreuteria paniculata, Acer creticum, Planera Richardi, Diospyros Lotus, Quercus Gramuntia, Quercus Ilex*, and *Pyrus Sorbus*.

Natural Orders on Plot E.

Verbenaceæ.
Labiatæ.
Plumbaginaceæ.
Plantaginaceæ.
Solanaceæ.
Polemoniaceæ.
Convolvulaceæ.
Boraginaceæ.
Celastraceæ (shrubby plants).
Leguminosæ.

Rosaceæ.
Sanguisorbaceæ.
Lythraceæ.
Saxifragaceæ.
Cucurbitaceæ.
Ranunculaceæ.
Berberidaceæ (shrubby plants).
Papaveraceæ.
Geraniaceæ.
Hypericaceæ (shrubby plants).
Malvaceæ.
Caryophyllaceæ.
Cruciferæ.
Cistaceæ (shrubby plants).
Rutaceæ.
Anacardiaceæ.
Ericaceæ (chiefly *Rhododendron*).

The most remarkable trees, climbers, &c., are *Periploca græca, Rhamnus catharticus, Prunus virginiana, Ostrya vulgaris, Ficus Carica, Morus alba, Cydonia vulgaris,* and *Mespilus germanica.*

Natural Orders on Plot F.

Polygonaceæ.
Chenopodiaceæ.
Nyctaginaceæ.
Phytolaccaceæ.
Urticaceæ.
Euphorbiaceæ.

On this plot stands the old Cedar of Lebanon (*Cedrus Libani*) one of the first plants introduced to this country. Two young trees have been planted on Plots E and F, right and left of the central walk from the Embankment. Here also grow *Quercus Ilex, Corylus Colurna, Broussonetia papyrifera, Ailantus glandulosa, Pistacia Terebinthus,* and *Fraxinus excelsior heterophylla.*

Natural Orders on Plot G.

Commelynaceæ.
Liliaceæ.
Melanthaceæ.
Dioscoreaceæ.
Smilaceæ.
Araceæ.
Amaryllidaceæ.
Iridaceæ.
Graminaceæ.
Cyperaceæ.

A notable specimen of *Quercus Ilex,* with a fine clean stem, stands on this Plot.

GILBERT AND RIVINGTON, PRINTERS, ST. JOHN'S SQUARE, LONDON.

Printed in the United States
By Bookmasters